The Burnout Companion
to Study and Practice

The Burnout Companion to Study and Practice: A Critical Analysis

WILMAR SCHAUFELI

Department of Psychology, Utrecht University, The Netherlands

AND

DIRK ENZMANN

Criminal Research Institute of Lower Saxony, Hannover, Germany

UK   Taylor & Francis Ltd, 1 Gunpowder Square, London EC4A 3DE
USA   Taylor & Francis Inc., 325 Chestnut Street, 8th Floor, Philadelphia,
        PA 19106

Adapted from *Opgebrand*, published by AD. Donker-Rotterdam,
by kind permission of the publisher

British Library Cataloguing-in-Publication Data

A catalogue record for this book is available from the British Library.

ISBN 0–7484–0697–2 (hardback)
ISBN 0–7484–0698–0 (paperback)

Library of Congress Cataloging-Publication-Data are available

Cover design by Youngs Design in Production

Typeset in Times 10/12pt by Graphicraft Limited, Hong Kong

Printed by T.J. International Ltd, Padstow, UK

Author biographies

Wilmar Schaufeli, PhD, is full Professor of Clinical and Organisational Psychology at Utrecht University, The Netherlands. He received his MA and PhD (*magna cum laude*) both in psychology from Groningen University, The Netherlands. He is a leading researcher in the field and has previously edited *Professional Burnout: Recent developments in theory and research* (Washington, DC: Taylor & Francis, 1993) with Christina Maslach and Tadeusz Marek. In addition to over 100 publications about unemployment, job stress, professional burnout, and absenteeism, he is actively involved in organisational consultancy and in psychotherapeutic treatment of burned-out employees.

Dirk Enzmann, PhD, graduated as a psychologist from the Psychological Institute of the Free University of Berlin, Germany. He taught educational psychology at the Technical University of Berlin and worked as a research assistant at the Social Educational Institute of Berlin. After receiving a PhD (*summa cum laude*) in psychology from the Free University of Berlin, he was employed at Utrecht University, The Netherlands, where he taught courses in research methods. Currently he is working as a senior researcher at the Criminological Research Institute of Lower Saxony, in Hannover, Germany. He is an active researcher on burnout, has conducted several burnout workshops, and (co-)authored three books and numerous journal articles on job stress and burnout. His current research interests are in health psychology and multivariate statistics.

Contents

Foreword

You are about to embark on a journey of discovery – the ongoing process of uncovering new information and new insights about something that is old and familiar. This is always an exciting journey for theorists and researchers, but it will be just as rewarding for clinicians and practitioners in organisational life. You will get a thorough, first-hand view of what it has been like to understand the dynamics of a compelling, 'grass-roots' phenomenon that has had a significant impact in the workplace.

The phenomenon is **burnout** – and everyone, it seems, already 'knows' what that is. Burnout has had global recognition in many workplaces, in many countries over many years. It has been the subject of countless books, articles and workshops – so much so that critics regularly make jokes about being 'burned-out on burnout.'

But on this score, the critics are wrong. When they, and you, read this book, the result will be a new view of burnout – a view that is more rich and complex in its breadth than its predecessors. Although this is not the very first book to be written about burnout, it is the first that tries to integrate and weave together *all* of these prior perspectives into a comprehensive and coherent whole. Rather than advocating for one particular position, the authors have tried to build on the strengths of all these varied contributions.

As one of the participants in this journey of understanding burnout, I have a special perspective on what Wilmar and Dirk have accomplished in this book. It began in 1990 – the year that I first met both of them in person. That year was a remarkable turning point in the field of burnout, for two reasons. It was the year that the first comprehensive international bibliography of burnout research was published – a bibliography that was co-authored by Dirk. It was also the year of the first international conference on professional burnout (held in Krakow, Poland). Wilmar was the co-organiser of this remarkable event, which brought together all the leading theorists and researchers on burnout to discuss and debate their different points of view on the phenomenon. Thus, nearly a decade ago, both Wilmar and

Dirk were the leading scholars working toward a synthesis of the field. It comes as no surprise, then, that they have been able to achieve this goal so successfully within *The Burnout Companion to Study and Practice*.

Their book is truly a companion in the sense of being a handbook or a guide for the entire field. If you want to know what has ever been said or studied about burnout, and how it has been evaluated, then this is the one true source to which you should refer. Indeed, if you have any interest whatsoever in burnout, this book is destined to occupy a prime spot on your bookshelf.

But for me, the book is also a 'companion' in the sense of being a helpful colleague and friend – and the authors have been just that, for many years, for both myself and many other people in the burnout field. It has been a remarkable journey for all of us to uncover the truths about burnout, and we are lucky that Wilmar and Dirk have done such as excellent job keeping track of all our various paths toward that ultimate understanding.

Enjoy the trip!

Christina Maslach
Berkeley, May 1998

CHAPTER ONE

Where does burnout come from?

History and background

Burnout is a metaphor. It describes a state of exhaustion similar to the smothering of a fire or the extinguishing of a candle. Where there used to be a vital spark and the flame of life was burning bright, it is now dark and chilly. The fuel has been used up and the energy backup is depleted.

As a matter of fact, this metaphor is not entirely appropriate as it is too static. A broken car battery that cannot be recharged and loses its power bit by bit is a better comparison than the notion of a fixed amount of energy that is slowly consumed until nothing is left, as in the case of a candle or of a fire. Indeed, burned-out individuals often describe themselves spontaneously as empty batteries that can no longer be recharged. This analogy to an empty car battery reflects the gradual process in which more energy has been consumed than was produced over a long period of time. A car battery gets empty when more power is used up than is supplied by its dynamo. In other words, the dynamic equilibrium between demand and supply of energy is disturbed and as a result the individual's energy backup is gradually exhausted. That is exactly what happens in burnout: people give too much for too long and receive too little in return. Essentially, burnout is the long term result of an imbalance between investments and outcomes.

Unfortunately, we have no single word which neatly catches this dynamic process by referring to the slow draining of energy from a battery. Instead, 'burnout' became the colloquial term for mental exhaustion, despite the false connotation of a fixed amount of energy that is drained to the dregs. Obviously, burnout is a very powerful metaphor because not only has it spread around very quickly in English speaking countries, but it is used almost universally. Although native equivalents exist in most Germanic languages, the English word 'burnout' is usually preferred.[1]

This chapter starts out with a brief history of the burnout construct. Next, the broader social context of occupational stress and burnout is discussed. Evidence is presented for the increase of stress and burnout in today's society. In the concluding section several explanations are offered for this increase.

1.1 THE RECENT HISTORY OF AN ANCIENT PROBLEM

The dictionary defines the verb 'to burn out' as 'to fail, wear out, or become exhausted by making excessive demands on energy, strength, or resources'. If we accept this broad definition for the time being, it is clear that burnout is not restricted to a particular historical period. Indeed, it was observed long before the notion itself was introduced within psychology in the early 70s. Probably the earliest example of burnout *avant-la-lettre* is found in a collection of poems attributed to William Shakespeare and published in 1599 as *The Passionate Pilgrim*:

> She burnt with love, as straw with fire flameth.
> She burnt out love, as soon as straw out burneth.   (Enzmann and Kleiber, 1989, p. 18)

At the turn of the century 'to burn oneself out' was borrowed from English slang meaning 'to work too hard and die early' (Partridge, 1961). Quite interestingly, the Japanese expression *karoshi* has an identical meaning – death by overwork – and can be considered the ultimate consequence of burnout: a lethal depletion of energy (Haratani, 1997).

More recently, but long before the discovery of psychological burnout, persons who suffered from it have been portrayed in great detail. For instance, in 1961 Graham Greene wrote a novel titled *A Burnt-out Case* which tells the sad story of the world famous architect Querry. Despite the fact that the novel was a bestseller, it did not make burnout a popular term in the 1960s (Box 1.1).

The best example of burnout *avant-la-lettre* is the case-study of a psychiatric nurse – Miss Jones – that was published in *Psychiatry* (Schwartz and Will, 1953). For nearly two decades, this description of burnout remained unprecedented. As is illustrated in Box 1.2, Miss Jones exhibits practically all the symptoms that were later to be identified as the typical elements of burnout: emotional exhaustion, a callous and indifferent attitude towards patients, and feelings of diminished accomplishment.

In the past, many other alternative notions have been used to describe a similar mental state as that captured by the burnout metaphor, for instance, 'overstrain' (Breay, 1913), 'neurocirculatory asthenia' (Cohen and White, 1951), 'surmenage' (Tuke, 1882), and 'industrial fatigue' (Park, 1934). Typically, these terms refer to negative aspects of the job that are considered to be the major contributors to the worker's poor mental condition. However, none of these words became as widespread and as popular as the term burnout that emerged in the 1970s.

The 'discovery' of burnout

Burnout was first mentioned as a psychological phenomenon that occurred in the helping professions by Bradley (1969), who proposed a new organisational structure in order to counteract 'staff burnout' among probation officers. However, the American psychiatrist Herbert Freudenberger is generally considered to be the originator of the burnout syndrome. In his influential 1974 paper entitled 'Staff burn-out' he describes the syndrome in great detail, thereby setting the stage for its introduction.

Box 1.1   Querry, the architect: A burnt out case

The world famous New York architect, Querry, 'is the victim of a terrible attack of indifference: he no longer finds meaning in art or pleasure in life' (back cover). Querry's cynicism is obvious when he states in a conversation with a friend 'To build a church when you don't believe in a god seems a little indecent, doesn't it? When I discovered I was doing that, I accepted a commission for a city hall, but I didn't believe in politics either. You never saw such an absurd box of concrete and glass as landed on the poor city square' (p. 114). Totally disenchanted Querry flees – anonymously – to the Congolese jungle, far away from everything and everybody that reminds him of his past. He is desperately looking for peace and reflection and ends up in a leper colony where he makes friends with Dr. Collins, the head of the colony. He confesses to him in their first conversation: 'Self-expression is a hard and selfish thing. It eats everything, even the self. At the end you find you haven't even got a self to express. I have no interest in anything anymore, doctor. I don't want to sleep with a woman, nor design a building' (p. 46). Dr. Collins diagnoses Querry as the mental equivalent of a 'burnt-out case'. They are 'lepers who lose everything that can be eaten away before they are cured' (p. 110). After having spent a while in the colony, Querry recovers and starts working again, this time for the lepers, until he discovers who he really is . . .

Graham Greene's fictional character is a gloomy, cynical, and disillusioned man who is at the end of the rope, no longer able any more to continue the work he once practised so passionately. In short, his creative energy is depleted: 'I began as an architect and I am ending as a builder. There is little pleasure in that kind of progress'.

Source: Graham Greene (1961, p. 110)

Freudenberger was employed as an unpaid psychiatrist in an alternative health care agency in New York. This *Free Clinic* was staffed mainly by young idealistic volunteers who were highly committed to their work with young drug addicts. Freudenberger observed that many volunteers experienced a gradual energy depletion and loss of motivation and commitment, which was accompanied by a wide array of mental and physical symptoms (Box 1.3).

To denote this particular state of exhaustion that usually occurred about one year after the volunteers started working in the *Free Clinic*, Freudenberger chose a word that was being used colloquially to refer to the effects of chronic drug abuse: 'burnout'. His detailed description of the burnout syndrome was easily and readily recognised, particularly by those who worked in the human services. In fact, Freudenberger's paper sparked the interest in burnout and constituted the basis for its future popularity. Quite interestingly, Freudenberger himself fell victim to burnout twice. It is therefore not surprising that he confesses:

Box 1.2 Miss Jones: The demoralised psychiatric nurse

Miss Jones worked as a nurse on a chronic ward in a psychiatric hospital. Most of her colleagues were well motivated and involved in their work with psychiatric patients. However, when Miss Jones returned after a period of absence she found the ward in a disconsolate state: the head nurse had been replaced, some nurses were absent, and the patients were restless. The nurses felt quite dissatisfied and their morale was poor. Miss Jones' colleagues reacted in an indifferent or even hostile way to her suggestions for changing the depressing situation. She felt rejected and frustrated and turned away from her colleagues. Now, she focused exclusively on the patients and cared for them with even more vigour than before. Meanwhile, the patients were more and more difficult to handle because of the poor treatment from the nursing staff and their fear of the impending chaos. Accordingly, Miss Jones had to pay the price for the indifference of her colleagues. This was quite painful for her since she expected gratitude and understanding from the patients, since she was the only nurse who really cared for them. After some time Miss Jones started to realise that she had failed. According to Schwartz and Will (1953) Miss Jones was now caught in a vicious circle: the more depressed and disappointed she felt, the more she failed, which in turn nourished her depressed mood and disappointment. Her depressed mood got worse, she felt exhausted and became more callous and indifferent, particularly towards her patients. Now, Miss Jones only saw the negative sides of her job, and she withdrew from social contact with patients and staff. Her job seemed meaningless to her and she started to be absent regularly. At this point the authors intervened by offering Miss Jones a couple of non-directive counselling sessions, which she readily accepted. In the first sessions she blew off steam. At a later stage, the authors taught Miss Jones how to analyse and understand her situation and how to lower her high expectations. The counselling approach succeeded. Little by little her situation improved now that her expectations had become more realistic. She got along better with her colleagues and her patients, which boosted her self-esteem. The case-study had a happy ending: some months after the intervention of the authors Miss Jones was doing well again.

Source: Schwartz and Will (1953)

> One of the prime motivators of my probing examination of Burn-Out was my own experience with it many years ago.   (Freudenberger and Richelson, 1980, p. xvii)

Independently, and almost simultaneously, Christina Maslach – a social psychological researcher – stumbled across the term burnout in California. She studied the ways in which people in stressful jobs cope with their emotional arousal and was particularly interested in cognitive strategies such as detached concern and dehumanisation used in self-defence. Maslach mainly interviewed health care workers like

Box 1.3   The first description of burnout

'For one, there is a feeling of exhaustion, being unable to shake a lingering cold, suffering from frequent headaches and gastrointestinal disturbances, sleepless-ness and shortness of breath. In short, one becomes too somatically involved with one's bodily functions. A staff member's quickness to anger and his instan-taneous irritation and frustration responses are the signs. The burn-out candidate finds it just too difficult to hold in feelings. He cries too easily, the slightest pressure makes him feel overburdened and he yells and screams. With the ease of anger may come a suspicious attitude, a kind of suspicion and paranoia. The victim begins to feel that just about everyone is out to screw him, including other staff members. The paranoid state may also lead to a feeling of omnipotence. The burning out person may now believe that since he has been through it all, in the clinic, he can take chances that others can't. He becomes overconfident and in the process may look foolish to all. His risk-taking behaviour in counsel-ling with speed freaks, psychotics, homicidal people and other paranoids some-times borders on the lunatic, in terms of his own behaviour. He may resort to an excessive use of tranquillisers and barbiturates. Or get into pot and hash quite heavily. He does this with the "self con" that he needs the rest and is doing it to relax himself. As to the person's thinking, that almost becomes a closed book. He becomes excessively rigid, stubborn, and inflexible. He almost cannot be reasoned with – he once again knows it all better than anyone else. He blocks progress and constructive change. Why? Because change means another adapta-tion and he is just too tired to go through more changes. Another behavioural indicator of burn-out is the totally negative attitude that gets verbalised. He becomes the "house cynic". Anything that is suggested is bad rapped or bad mouthed. He knows it all because he has been through it all. The person looks, acts and seems depressed. He seems to keep to himself more. Other brothers and sisters really don't know what is going on. But they do know that changes are taking place in that individual. A sign that is difficult to spot until a closer look is taken is the amount of time a person is now spending in the free clinic. A greater and greater number of physical hours are spent there, but less and less is being accomplished. He just seems to hang around and act as if he has nowhere else to go. Often, sadly, he really does not have anywhere else to go, because in his heavy involvement in the clinic, he has just about lost most of his friends.'

Source: Freudenberger (1974, pp. 160–161). Quoted with permission from Blackwell Publishers.

physicians, nurses, psychiatrists, and hospice counsellors. Three general themes emerged from these interviews (Maslach, 1993). First, many practitioners talked about being emotionally exhausted and drained of all feeling. Second, the inter-viewees developed negative perceptions and feelings about their patients. Finally, all too often the practitioners experienced a crisis in professional competence as a result of the emotional turmoil. When, by chance, she described these results to an attorney, she was told that poverty lawyers called this particular phenomenon 'burn-out'. Once Maslach and her colleagues had adopted this term, they discovered that the metaphor was immediately recognised by their respondents: a new psychological notion was born.

Approaches to burnout

The discovery of burnout, almost simultaneously in the early 70s at America's east and west coasts by Herbert Freudenberger and by Christina Maslach and her colleagues, illustrates that burnout first emerged as a social problem and not as a scholarly construct. Apparently, they had discovered something that was 'in the air'. This social origin of burnout is essential for understanding the historical development of the concept that took place along two lines that are relatively separate.

The clinical approach

Initially, the focus was predominantly on elaborate clinical descriptions of the burnout-syndrome (see Box 1.3). Attempts were made to characterise burnout by means of cautious but unstandardised observations and individual case studies. In particular, attention was paid to the symptoms that are displayed by burned-out individuals. Following the medical tradition, these symptoms were then grouped in order to identify a syndrome, i.e. a set of symptoms that occur together and consti-tute a recognisable negative condition. For instance, Freudenberger (1974) described not only physical (e.g. headaches) and behavioural signs of burnout (e.g. use of illicit drugs), but also affective (e.g. depressed mood), cognitive (e.g. cynicism), and motivational symptoms (e.g. demoralisation). The clinical approach emphasises the importance of individual factors underlying the burnout-syndrome.

Typically, the first publications on burnout were anecdotal and appeared in journals, magazines, and periodicals not only for professionals such as teachers, social workers, nurses, physicians, and managers, but also for pharmacists, fire-fighters, and librar-ians. Stimulated by these publications, public interest grew enormously and burnout became a hot topic. In further popularising the subject, the mass media played a crucial, albeit debatable, role. That is, burnout ended up as the buzzword or catch-phrase of the late 70s and early 80s. Meanwhile, the concept was stretched to encompass far more than it did originally. This blurred, all-encompassing meaning of burnout and the unempirical nature of the clinical approach led academic critics to disparage the concept of burnout or even dismiss it entirely. Initially, the popularity of burnout was not surprisingly inversely proportional to the scholarly interest it evoked.

Although today the clinical approach is no longer as dominant as it was initially it survived remarkably well. There is still a stream of publications that are typically clinical in nature and that are written by practitioners who work with burned-out professionals. For instance, recently Figley (1995) edited a book on 'compassion fatigue' that focused on secondary traumatic stress among professionals who deal with severely traumatised persons. Interestingly, the mainly descriptive clinical approach to burnout developed independently from the more analytical research approach to burnout.

The research approach

Social psychologists such as Christina Maslach and Ayala Pines placed burnout on the scientific agenda against all the odds. Soon after they had come across the syndrome in a number of interviews they initiated a research programme. An important development came from the introduction of short and easy to administer self-report questionnaires to assess burnout (see Chapter 3). Initially the predominantly negative attitude of academics towards burnout seriously hampered their empirical work. As Maslach and Jackson (1984a, p. 139) put it:

> Because it has a catchy ring to it, burnout is sometimes immediately dismissed as fad or pseudoscientific jargon that is all surface and no substance.

For instance, their joint psychometric article on the development of the Maslach Burnout Inventory (MBI) was returned by a scientific journal editor with a short note that it had not even been read 'because we do not publish "pop" psychology' (Maslach and Jackson, 1984a, p. 139). Ironically, the MBI is now the most widely used and best validated instrument to measure burnout (see Chapter 3).

As a result of the development of standardised inventories, burnout could now be studied empirically. The prevailing conceptual confusion about the nature of burnout was gradually clarified by the almost general acceptance of the MBI as the main instrument to assess the syndrome. By implication, the definition of burnout used by the test-authors was sanctioned by the research community. At first Maslach and Pines and their collaborators studied health care professionals, but soon they extended their research to other occupational groups within the human services including teachers, social workers, police officers, and prison officers. They chose to study these human services professionals because of the high emotional demands they supposedly experienced in their work. Contrary to the clinical approach that stresses the importance of individual factors, the first social-psychological research emphasised the interpersonal nature of burnout:

> Despite similar noxious effects as other stress reactions, the unique feature of burnout is that its stress results from the *social* interactions between helpers and their recipients. (Maslach, 1982b, p. 3)

At about the same time, other researchers like Cary Cherniss (1980a, 1980b) and Robert Golembiewski and his colleagues (Golembiewski *et al.*, 1986) stressed the importance of the organisational environment in the development of burnout. Organisational psychological research investigated burnout in a way that touched the concern, not only of the scientific community, but also of administrators, managers, policy-makers, and organisational consultants. Although some authors, like

Cherniss, argued that burnout was related to organisational factors that were typically found in human services organisations, others, like Golembiewski, rejected this restriction. It seems that recently the tendency to associate burnout exclusively with human services professionals is weakening and the scope has broadened to other fields and professions (Maslach and Leiter, 1997). In the early 90s burnout research entered a new phase in which theoretically-driven and methodologically-sound studies are increasingly conducted.

1.2 OCCUPATIONAL STRESS (BURNOUT) AS A SOCIAL PROBLEM

In order to place burnout in a broader societal and cultural context we must broaden our scope somewhat by including occupational stress. Occupational stress is a more generic term that refers to any affect-laden negative experience that is caused by an imbalance between job demands and the response capability of the worker. When job demands are too high to cope with, stress reactions are likely to occur. Burnout is considered to be a special type of prolonged occupational stress that results particularly from interpersonal demands at work. Traditionally, occupational stress research is carried out predominantly in industrial settings, thereby neglecting the human services, whereas the reverse is true for burnout. In the next chapter, the relationship between burnout and occupational stress is discussed in greater detail. In the remainder of this chapter we will argue that occupational stress, including burnout, is a growing problem in western societies and we will suggest some factors that can be held responsible for this marked increase.

Prevalence of occupational stress (burnout)

There is ample evidence that the prevalence of occupational stress is rather high. For instance, a survey of nearly 16 000 workers in all 15 European Union member states sponsored by the European Commission revealed that 29% considered that their work activity affected their health (Paoli, 1997). The work-related health problems mentioned most frequently were back-pain (30%), stress (28%), and overall fatigue (20%). A similar survey in the United States showed that more than 75% of the American workers '. . . describe their jobs as stressful and believe that the pressure is steadily increasing' (International Labour Office, 1993, p. 65). Furthermore, US industry loses approximately 550 million working days per year due to absenteeism, of which 54% is estimated to be in some way stress-related (Elkin and Rosch, 1990). It has been estimated that in the United Kingdom 30 to 40% of all sickness absence is attributable to some form of mental illness (O'Leary, 1993). In the United Kingdom, a National Survey of Health and Development of almost 1500 young men showed that 38% of the sample were under some or severe 'nervous strain' at work, whereas only 8% were under similar strain at home or in their personal lives (Cherry, 1978). The top three causes reported were: pressure of work (36%), responsibility (24%), and contact with people (12%). The last cause is particularly remarkable because, as noted before, demanding interpersonal relations are considered to be a specific antecedent of burnout. Prevalence studies in human services professions yield perhaps even more dramatic results. In a study among nearly 750 psychologists, Guy *et al.* (1989) found that 74% reported 'personal distress' during the previous three years, and 37% of this group believed that their distress decreased the quality of care they provided to their clients. In a similar vein, 62% of the members of Division 29 (Psychotherapy) of the American Psychological Association admitted to continuing working when too distressed to be effective (Pope *et al.*, 1995).

Prevalence rates of occupational stress are not only high but they are also rising continuously. For instance, Sutherland and Cooper (1990) reported that the number of workdays lost in the United Kingdom for stress-related causes increased significantly between 1982 and 1985. Of the 32.8 million days lost in 1984–5, 5.53 million (16%) were due to mental health causes. In the United States, Northwestern National Life Insurance Company (1991) reported that in 1991 25% of their sample had stress-related illnesses compared with 13% in 1985. Forty-six per cent of these workers felt highly stressed compared with 20% 6 years earlier. An increase of stress-related absenteeism can also be observed over a fairly long period. Hingley and Cooper (1986) calculated that in Britain absenteeism due to 'nervousness, debility and headache' has increased by 528% in a 25 year period from 1955 to 1979!

Workers' disability rates

In most industrialised countries the number of disability benefit recipients has increased sharply over the past two decades. Prins (1990) compared Belgium, Germany, and the Netherlands and showed that the steepest rise was observed in the Netherlands. Most interestingly, in all three West European countries, mental disorders were the second major diagnostic group after musculoskeletal diseases. Figure 1.1 shows that in the Netherlands in 1967 (the year of the introduction of the Disability Security Act) mental disorders accounted for 11% of the disability benefit recipients. This rate continued to rise by about 10% each decade but seems to have remained stable at about 30% since the early 90s. Most probably, these figures are underestimates since diseases with known psychosomatic aetiologies such as lower back pain or myocardial infarction are not considered to be mental disorders – they are recorded instead as musculoskeletal and coronary heart diseases, respectively.

As can be seen in Figure 1.1, work incapacity due to musculoskeletal disorders rose only slowly, whereas work incapacity due to coronary heart disease dropped sharply. At present in the Netherlands about one in every three disability benefit recipients is assessed work disabled on mental grounds. This means that approximately 34 000 Dutch workers receive a diagnosis of this kind – 116 people each day, one person every 4 minutes in the working day. Among human services professionals these figures are even more dramatic. In the Netherlands, these professionals are civil servants and have their own disability security arrangements. They

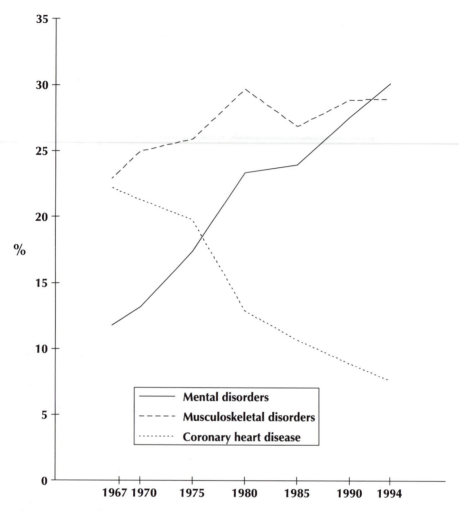

Figure 1.1   Workers' disablement in the Netherlands (1967–1994): Relative frequency of the three main diagnostic groups. *Source*: Statistical Yearbooks of the Gemeenschappelijk Medische Dienst (GMD) 1968–1995. Amsterdam: GMD.

are therefore not included in Figure 1.1. Among work incapacitated civil servants, 47% (about 4500 cases annually) receive benefits on mental grounds. The highest rate is observed in teachers (60%).

A closer inspection of these mental cases reveals an interesting profile. The vast majority (84%) suffer from an adjustment disorder and are diagnostically labelled as having a 'situative or exogenous reaction'. Moreover, recipients who receive disablement benefits on mental grounds are relatively young (aged under 35), predominantly highly educated, and work mostly in service professions such as health care, education and social work (Kers and Van der Zouwe, 1994). As we will see in Chapter 4, this profile closely matches the typical burned out professional.

In the United States stress-related workers' compensation claims have also risen sharply from less than 5% in 1980 to 15% in 1989 (International Labour Office, 1993). The state of California has the highest rate with 17%. In addition, the California Workers' Compensation Institute (CWCI, 1990) reports that, during the 10 year period from 1979 to 1988, the frequency of mental health claims for every 1000 covered workers increased by 540%, whereas the incidence of all disabling injuries declined by 8%. As in the Netherlands, the highest incidence of stress claims originated from public sector employees (e.g. law enforcement personnel and teachers). In 1986, there were 1.7 stress claims filed for every 1000 public sector employees, which is nearly six times the frequency for employees in the private sector. Based on these and similar data the CWCI concluded that 'mental stress claims are a significant – and growing – segment of the Californian worker's compensation program' (p. 6).

Costs

The expenditure on occupational stress is huge. It is estimated that in the United States occupational stress costs employers in excess of $200 billion per year in absenteeism, reduced productivity, medical expenses, and compensation claims (International Labour Office, 1993). Moreover, stress claims are relatively expensive: the average cost of a stress claim to the largest underwriter of workers' compensation insurance in the USA (the Liberty Mutual Group) in 1993 was $13 000, while the average cost of all claims reported was $6000 (Brogmus, 1996). Thus stress claims are twice as expensive as other claims. In the United Kingdom, the cost of sickness absence for stress and mental disorders have been estimated at more than £5 billion per year, which amounts to over 10% of the Gross National Product (Cooper *et al.*, 1996). In the Netherlands in 1994 about half of the costs for absenteeism and workers' compensation claims, which amount to approximately NLG 50 billion per year, are spent on stress-related disorders. Workers' compensation claims alone, of which about 30% are mental in nature, accounted for 13.9% of the Dutch Gross National Product in 1994. This is the highest percentage in the European Union (average = 9.6%).

Similar trends concerning the growing incidence of occupational stress have been observed in not only other industrialised countries such as Australia, Canada, Germany and Japan, but also in fast developing countries such as India and Brazil. This has led the International Labour Office (1992, p. 4) to speak about the 'globalisation of stress'. Moreover, it concludes that 'occupational stress is an important concern for workers, enterprises and society' (p. 15) and that 'the economic impact of stress on society is large and growing' (p. 15).

In summary, occupational stress is an important and rising concern in industrialised countries, particularly in the human services. It seems that the level of occupational stress has increased alarmingly in previous decades. This is illustrated by increasing stress-related absenteeism and work incapacity rates as well as by the rising associated costs. A closer examination of these figures suggests that burnout might play a substantial role.

1.3 WHY DOES OCCUPATIONAL STRESS (BURNOUT) INCREASE?

In order to understand the apparent increase of occupational stress and the sudden appearance and immense popularity of burnout we have to take into account not only work-related factors, but also the changing social, cultural, and ideological context. Seven tentative factors are considered that might explain the widespread prevalence of occupational stress and that might have set the stage for burnout. Of course, these factors are interrelated and do not operate independently from each other. In one way or another all the factors are aspects of a global economic, social and cultural transformation process that has affected society as a whole.

1. The emergence of the service sector

In recent decades, in all industrialised countries the commercial and not-for-profit service sector has rapidly grown at the expense of more traditional sectors such as agriculture and particularly manufacturing. In the European Union, 59% of the workforce is employed in the service sector, against 7% in agriculture and 34% in industry (Paoli, 1997). The relative size of the service sector varies from 69% in the Netherlands to 50% in Portugal. The human services constitute an important and growing part of the service sector.

Historically speaking, the human services became professions for the first time in the 1920s in the United States as well as in most European countries. At that time, the human services were small and uninfluenced by government policy or bureaucracy. This changed radically after World War II when human services work rapidly became professionalised, bureaucratised, and accredited. Government interference, support, and control increased since the human services were considered more and more an instrument to solve the problems of society. For instance, president Kennedy's *War against Poverty* in the 1960s boosted the numbers of human services professionals. During recent decades the number of people employed in services professions has risen sharply. Professions such as nursing, teaching, and social work are among the fastest growing occupational groups in the United States (Rifkin, 1995). For instance, in just 4 years, the number of nurses has increased by 14%. Because of this sharp increase in services professionals a growing number of employees are at risk of burning out. As we have seen previously, employees in service professions run a particularly high risk of developing burnout because of the emotional demands they are facing in working daily with people. Thus, the fact that occupational stress and burnout have increased might, at least partly, be explained by the growth of the service sector in general and the human services in particular.

2. Labelling

These days people are more inclined to label their problems, concerns, complaints, ailments, troubles, and difficulties in psychological terms than they used to do some decades ago. Stress plays a pivotal role in this labelling process, since it is considered the root cause of many symptoms. Abbott (1990) argues that the term stress is particularly suited for labelling purposes since it is ambiguous, ambivalent, and over-inclusive. During the last decades, the use of the word stress has spread from scientific and professional circles to the general public. Barley and Knight (1992) showed that academic interest in stress not only began earlier (in the 1950s), but also popular interest did not set in until after the first surge of research subsided (in the 1970s). Furthermore, they argued that new professions like nursing and social work have used the rhetoric of stress as an organisational vocabulary.

> Claims of stress may be used to galvanise a sense of consciousness and solidarity among an occupation's members. Moreover, by repeatedly proclaiming exposure to stressful work, an occupation may construct a publicly credible rationale for why it should be allowed such privileges as higher pay and the right to self-regulate. (Barley and Knight, 1992, p. 19)

Thus, in recent decades, the notion of 'stress' has become most important in labelling various negative individual experiences, even up to the point that it developed into a cultural symbol of our time. Academic research and the rapid growth of new (human services) professions stimulated the diffusion of stress within our society. It is likely that similar dynamics play a role in the increased popularity of burnout.

3. Individualisation

In modern society, social roles are no longer fixed since the social fabric of traditional communities (e.g. church, neighbourhood) has gradually eroded. Instead, people have to define their roles themselves, and build and maintain their own social networks. This requires considerable effort and social skill. An increasing number of individuals do not have at their disposal the necessary psychosocial resources and 'move with dizzying speed toward greater anonymity, impersonality, and disconnectedness' (Slater, 1976, p. 128). According to Lasch (1979), we are living in a narcissistic culture that is characterised by the increasingly transient, unrewarding, and even combative nature of social relationships. As a result, the development of narcissistic, self-absorbed, manipulative individuals who demand immediate gratification of their desires but remain perpetually unsatisfied is fostered. Hence, individualisation leads to greater stress (i.e. alienation, disconnectedness), as well as to fewer resources to cope with frustration (i.e. narcissism). The combination of these two trends produce 'a perfect recipe for burnout' (Farber, 1983, p. 11).

4. Increased mental and emotional workload

New technology has penetrated not only manufacturing, but also the service sector. As a consequence, in many occupations a shift has been observed from a physical to a mental workload. For instance, a nurse in an Intensive Care Unit operates in a sophisticated and highly technological environment that requires the use of complex cognitive skills such as vigilance, accuracy, and rapid decision making. These cognitive demands that follow from the introduction of new technology increase the workers' mental workload and may contribute to burnout (Schaufeli *et al.*, 1995). In addition, a shift towards an increasing emotional workload can be observed. Because of fierce competitiveness, employees are forced to continuously display 'consumer friendly' attitudes, in ways that contradict the expression of their genuine feelings. An illustrative situation comes from the work of Hochschild (1983), who studied occupational stress of flight attendants suffering from the 'emotional labour' associated with the need to continually please passengers and maintain a permanent smile:

> A young businessman said to a flight attendant, 'Why aren't you smiling?' She put her tray back on the food cart . . . and said, 'I'll tell you what. You smile first, then I'll smile.' The businessman smiled at her. 'Good,' she replied. 'Now freeze, and hold that for fifteen hours.'   (Hochschild, 1983, p. 127)

But it is not just the qualitative workload that has increased in many occupations, it is also the quantitative workload. In the European Union, 18% of workers indicate that they are continually under time pressure, whereas 35% indicate that this happens regularly (at least 50% of the time) (Paoli, 1997). Houtman and Kompier (1995) analysed data from the National Work and Living Survey of a national representative sample of the Dutch labour force conducted in 1977, 1983, 1986, and 1989 by the Netherlands Central Bureau of Statistics. These surveys showed that the percentage of workers who complained about excessive work pace steadily increased from 38% in 1977 to 51% in 1989. This equals an increase of 13% in 12 years. Interestingly, those who work predominantly with other people such as managers, secretaries, health care workers (e.g. physicians and nurses), catering personnel (waiters, cooks, and barkeepers), and teachers experienced the highest work pace.

More particularly, Cherniss (1980a) points to the increasing case-load as a major cause of quantitative work overload contributing to professional burnout in the human services. Because of the individualisation of society, communities have declined and with it informal support systems. As a result, the service demands on the formal human service institutions that had replaced the old informal structures increased dramatically, particularly after World War II. Furthermore, Cherniss (1980a) argues that after the oil-crisis in the early 1970s governments in most countries started to economise and cut back public sector funds. Professionals working in public agencies faced in particular increases in service demands and declining budgets leading to increased workloads and hence more job stress. As Cherniss (1995, p. 36) puts it:

> Clients became needier and more difficult to help, and the numbers needing help increased. At the same time, support for the human services declined. Thus professionals had to do more with less.

5. The weakening of professional authority

Traditionally, professionals were appreciated members of society who owned considerable prestige and social status. However, during the 70s, the situation changed dramatically in such a way that 'the heroes of the 80s were not idealistic teachers or physicians; they were stockbrokers and corporate executives' (Cherniss, 1995, p. 35). Moreover, the fundamental assumptions on which authority was rested were no longer taken for granted and even the professionals' knowledge and skill were questioned. The general public started to mistrust professionals as well as the social institutions they represented. In a Dutch bestseller, published in the late 1970s, human services professionals were portrayed as manipulators who first consciously and willingly constitute a 'market of well-being and happiness' which they next monopolise (Achterhuis, 1979). The professionals were accused of misusing public funds since they continuously create new demands and service areas just in order to safeguard their own professional existence. A similar change was observed in the United States by Cherniss (1995, p. 35):

> Clients were . . . more resistant and suspicious as they came to believe that professionals were not there to help clients but to protect themselves.

These kinds of criticisms legitimatise governmental interference in a period in which public funds were limited anyway. Cherniss considers the growing disenchantment with the professionals and their declining authority as a major social cause of professional burnout.

The weakening of the professional authority was accompanied and reinforced by the entitlement of clients, patients, and customers. For instance, Gow (1982) argues that the public image of nursing care has changed considerably during the past two decades. A caring and committed nurse used to be responsible, orderly, tidy, neat, prudent, industrious, disciplined, and sensible. Nowadays, nurses are supposed to be empathic, giving, and in tune with the emotional lives of their patients. Accordingly as a result of patient entitlement the standards of care and compassion have risen, resulting in considerable emotional demands for the workers. In short 'we expect more of helping professionals than ever before' (Cherniss, 1995, p. 6).

6. 'Professional mystique'

According to Cherniss (1980a, pp. 249–256), 'professional mystique' is a set of beliefs, expectations, or opinions that the general public holds with regard to professionals and their work. These particular ideas are perpetuated by the mass media and transmitted in most professional training programmes. They reinforce unrealistic expectations, particularly to newcomers in the field. Inevitably, this mystique clashes with the harsh reality the professionals face, ultimately culminating in disillusionment and burnout. Cherniss distinguished between five different elements of the professional mystique:

- **Competence** All too often credentials are confused with competence. When professionals enter their first jobs they have the necessary credentials but generally do not feel like the competent expert that the public expects them to be.

- **Autonomy** Traditionally, professional status has been associated with freedom and control over the decisions affecting one's work. However, in practice, this turns out to be a thwarted expectation. More often than not bureaucratic interference, strict regulations, and fixed schedules are the standard.

- **Self-realisation** Professional jobs are supposed to be challenging, interesting, and varied, thus providing stimulation and fulfilment. However, in reality, professional practice is often dull, insignificant, and routine instead of being heroic or charismatic.

- **Collegiality** Many professionals like teachers work rather isolated from their colleagues. But even when frequent contacts with colleagues do occur they are often characterised by undercutting rivalry, competition, and distrust. So the expected collegiality is repeatedly frustrated.

- **Attitude of recipients** Typically, the recipient is supposed to be grateful, honest, and co-operative. Unfortunately, a disturbingly large percentage of recipients fail to live up to the professional's ideal for clients. Not only do recipients not show their gratitude, they sometimes tell lies, cheat, manipulate, or even actively resist the professional's effort to help.

7. Changed psychological contract

Due to changes in the world of labour and to changes in expectations of the workers, the nature of the so-called psychological contract between employee and organisation has changed over recent years. A psychological contract is defined as an individual's belief regarding the terms and conditions of a reciprocal exchange agreement between him/her and the organisation (s)he is working for (Rousseau, 1989). In other words, the psychological contract entails a belief in what the employer is obliged to provide based on perceived promises of reciprocal exchange. In recent years, this balance of exchange has shifted in a less favourable direction for the employee. Generally speaking, employees have to give more and receive less from their employer. For instance, as we have previously seen, the workload has increased qualitatively as well as quantitatively. On the other hand, life-time employment is no longer the norm. More and more workers are employed on a temporary or part-time basis or as contractors or freelancers – so-called contingency workers. It is illustrative that *Manpower*, the United States' largest temp agency, is now that country's single largest employer with 560 000 workers (Rifkin, 1995). Having to work harder and at the same time experiencing future job insecurity erodes the psychological contract. Violation of the psychological contract may result not only in negative consequences such as turnover and dissatisfaction (Robinson and Rousseau, 1994), but also in reduced organisational commitment and burnout (Schaufeli *et al.*, 1996b) and in coronary heart disease (Siegrist, 1996).

Violation of the psychological contract is all the more likely since today the expectations that are held about a job are higher then ever before (*cf.* the professional
mystique). The economic nature of the job ('earning money to make a living') is
supplemented by myriad other expectations hardly any job can live up to: good
promotion prospects, challenging work, nice colleagues, variety, significance, participation in decision making, and autonomy. Of course, these expectations are
legitimate and the quality of one's working life goes beyond the pay-cheque, but in
today's world of work, they are hard to get. Hence, these high expectations are
built-in sources of frustration and therefore increase the risk of burning out.

1.4 SUMMARY

Burnout is not a new phenomenon – it has its roots in the past. However, because of
a unique constellation of several factors it was 'discovered' in the early 70s as a
particular type of prolonged occupational stress that seemed to occur most promin-
ently among human services professionals. Figures from several countries indicate
that occupational stress, including burnout, has risen sharply in recent decades. This
is also true of the costs that are associated with this pressing social problem: most
countries of the European Union spend about 10% of their GNP on the negative
consequences of occupational stress such as absenteeism, sick-leave, and disability
claims. Seven factors are distinguished that may have contributed to the increase of
work-related stress. Four of them are more or less specific for burnout: the expan-
sion of the service sector; increased mental and emotional workload; the weakening
of the professional's authority and the entitlement of service recipients; a particular ideology that reinforces unrealistic expectations in professionals (*professional
mystique*). The remaining three factors are less specific and apply to the increase of
genuine occupational stress: labelling of problems in living as 'stress'; individualisation of society; the changed psychological contract with the employer (i.e. more
has to be performed for fewer rewards).

The development of burnout as a psychological notion took place along two
lines. Initially, in the pioneer phase a clinical approach prevailed that was characterised by merely describing the symptoms of the burnout syndrome. In the second,
empirical phase, social and organisational psychologists studied burnout more
systematically, using standardised instruments (mainly self-reports). Recently, more
theory-driven research on burnout has been conducted. In the next two chapters we
will elaborate on the clinical and the scientific approaches to burnout by discussing
its symptoms and by describing the ways that burnout is assessed, respectively.

Note

1   For instance, in Afrikaans ('uitbranding'), Dutch ('opgebrand'), German ('ausgebrannt'),
     Norwegian ('utbrenthet'), and Swedish ('utbränd').

CHAPTER TWO

What is it?

Symptoms and definitions

Myriad possible burnout symptoms and definitions exist. The more or less exhaustive list of burnout symptoms that is presented in the first part of this chapter shows that the concept can easily be expanded to mean anything, so that there is the danger that in the end it does not mean anything at all. Therefore, in the second part of this chapter we will discuss several definitions of burnout that emphasise the specific and distinctive nature of the syndrome. The chapter concludes with a section called 'old wine in new bottles?' that shows that burnout can be differentiated from related concepts like job stress, depression, and chronic fatigue.

Before discussing the various symptoms of burnout in greater detail let us first have a look at Mr Dijkstra (see Box 2.1), a teacher who had been referred by his occupational physician for psychotherapeutic treatment after being on sick-leave for over 6 months.

Mr Dijkstra was suffering from a broad range of burnout symptoms, which are quite easy to recognise with the previous preliminary descriptions of the syndrome in mind: physical and mental exhaustion, depressed mood, irritability, headaches, restlessness, and avoiding contact with other people. But, above all, he felt disappointed and powerless. In Chapter 6 we will return to Mr Dijkstra and see how his psychotherapy developed.

2.1 POSSIBLE SYMPTOMS OF BURNOUT: FROM ANXIETY TO LACK OF ZEAL

Table 2.1 displays 132 symptoms that have been associated with burnout – an *A* (anxiety) to *Z* (lack of zeal) of burnout! (*cf.* Beemsterboer and Baum, 1984; Burisch, 1989; Cordes and Dougerthy, 1993; Einsiedel and Tully, 1982; Freudenberger, 1974; Kahill, 1988; Maher, 1983; Paine, 1982).

Most of the symptoms displayed in Table 2.1 come from uncontrolled clinical observations or from interview studies with an impressionistic or unspecified analysis of data, rather than from rigorously designed and thoroughly conducted quantitative

Box 2.1   Mr Dijkstra, the disappointed teacher

During the past 2 years, Mr Dijkstra, a 48 year old teacher at a vocational training centre, has played a crucial role in the merging of his school with another training centre. It has been a very hectic and busy time because he was one of the advocates and active agents who promoted that merger. After the merger was concluded Mr Dijkstra felt very disappointed since he was not promoted to the newly created job as department coordinator in the fresh organisation. Instead, the job he hoped to receive was offered to a younger colleague who had always been sceptical of the merger. Mr Dijkstra felt hurt, resentful, and unfairly treated: in his opinion, he had put much more time and effort into reorganising the school than his younger colleague – yet he was denied the appropriate reward. Soon after this event, Mr Dijkstra developed particular symptoms. He had occasionally felt tired before, but now it was different – he felt completely mentally exhausted. It took an extreme effort to take on anything. In earlier times he used to quickly recuperate from his tired-ness after a weekend or a couple of days off. Since then he has been on sick-leave for over 6 months and was still unable to perform his job because he felt extremely tired and anxious. He slept till ten o'clock in the morning, needed an additional couple of hours to really wake up and nevertheless kept feeling tired all day long. Already some time ago – during that busy period at school – Mr Dijkstra had quit his hobbies – refurbishing antique furniture and playing bridge. Although he would have enough time to pursue these hobbies now, he lacked the energy and didn't fancy it. Instead, he worried a lot and had problems concentrating. For instance, when reading the newspaper, after reading some lines he was forgetting what he had read previously. Moreover, he suffered from headaches and pain in the neck, and felt depressed and restless. Mr Dijkstra was particularly uncomfortable in interpersonal situations: he felt distressed – his throat became tight and he was not able to breathe normally. As a result, he started to avoid social situations and became more and more isolated from colleagues, friends, neighbours, and relatives. If things did not work out properly or when somebody was unkind, Mr Dijkstra got upset. He was irritable, emotional, and easily hurt, which strained his family, especially his two teenaged children. But perhaps the most frightening of all – Mr Dijkstra didn't recognise himself anymore, he felt powerless and totally out of control. He could not understand what had happened to him.

Source: Schaap *et al.* (1995)

studies. Nevertheless, it is useful to consider all possible burnout symptoms briefly before focusing on the core symptoms since this will illustrate the initial broad and almost all-inclusive scope of the syndrome.

The categorisation used in Table 2.1 follows the common classification of psychological symptoms in five clusters: affective, cognitive, physical, behavioural, and

Table 2.1 Possible burnout symptoms
a) symptoms at individual level

affective	cognitive	physical	behavioural	motivational
depressed mood	helplessness	headaches	hyperactivity	loss of zeal
tearfulness	loss of meaning and hope	nausea	impulsivity	loss of idealism
emotional exhaustion	fear of 'going crazy'	dizziness	procrastination	disillusionment
changing moods	feelings of powerlessness and impotence	restlessness	increased consumption of: caffeine, tobacco, alcohol, tranquillisers, illicit drugs	resignation
decreased emotional control	feelings of being 'trapped'	nervous tics		disappointment
undefined fears	sense of failure	muscle pains	over- and undereating	boredom
increased tension	feelings of insufficiency	sexual problems	high risk-taking behaviours (e.g. skydiving)	demoralisation
anxiety	poor self-esteem	sleep disturbances (insomnia, nightmares, excessive sleeping)	increased accidents	
	self-preoccupation	sudden loss or gains of weight	abandonment of recreational activities	
	guilt	loss of appetite		
	suicidal ideas	shortness of breath	compulsive complaining	
	inability to concentrate	increased pre-menstrual tension		
	forgetfulness	missed menstrual cycles		
	difficulty with complex tasks	chronic fatigue		

Table 2.1 a) (Cont'd)

affective	cognitive	physical	behavioural	motivational
	rigidity and schematic thinking	physical exhaustion		
	difficulties in decision making	hyperventilation		
	daydreaming and fantasising	bodily weakness		
	intellectualisation	ulcers		
	loneliness	gastric-intestinal disorders		
	diminished frustration tolerance	coronary diseases		
		frequent and prolonged colds		
		flare-ups of pre-existing disorders (asthma, diabetes)		
		injuries from risk-taking behaviour		
		increased heart rate		
		high blood pressure		
		increased electrodermal response		
		high level of serum cholesterol		

Table 2.1 b) symptoms at interpersonal level

affective	cognitive	physical	behavioural	motivational
irritability	cynical and dehumanising perception of recipients		violent outbursts	loss of interest
being oversensitive	negativism with respect to recipients		propensity for violent and aggressive behaviour	discouragement
cool and unemotional	pessimism with respect to recipients		aggressiveness towards recipients	indifference with respect to recipients
lessened emotional empathy with recipients	lessened cognitive empathy with recipients		interpersonal, marital and family conflicts	using recipients to meet personal and social needs
increased anger	stereotyping of recipients		social isolation and withdrawal	overinvolvement
	labelling recipients in derogatory ways		detachment with respect to recipients	
	'blaming the victim'		responding to recipients in a mechanical manner	
	air of grandiosity		isolation or overbonding from other staff	
	air of righteousness		sick humour aimed at recipients	
	'martyrdom'		expression of hopelessness, helplessness and meaninglessness towards recipients	
	hostility		using distancing devices	
	suspicion		jealousy	
	projection		compartmentalisation	
	paranoia			

Table 2.1 c) symptoms at organisational level

affective	cognitive	physical	behavioural	motivational
job dissatisfaction	cynicism about work role		reduced effectiveness	loss of work motivation
	feelings of not being appreciated		poor work performance	resistance to go to work
	distrust in management, peers and supervisors		declined productivity	dampening of work initiative
			tardiness	low morale
			turnover	
			increased sick-leave	
			absenteeism	
			theft	
			resistance to change	
			being over-dependent on supervisors	
			frequent clock watching	
			'going by the book'	
			increased accidents	
			inability to organise	
			poor time management	

motivational. In addition, three levels are distinguished since burnout typically includes not only individual symptoms, but also symptoms at the interpersonal and organisational levels.

1. Affective symptoms

Typically, a gloomy, tearful and depressed mood is observed among those who suffer from burnout. Although moods may change quickly, generally spirits are low, and a sad and dim mood prevails. The individual's emotional resources are exhausted because too much energy has been used for too long a time. In addition, emotional control might be decreased which can lead to undefined fears, anxiety and nervous tension.

In interpersonal contacts with others the burned-out professional can be irritable and oversensitive but also cool and unemotional. The last is illustrated by the fact that lessened emotional empathy with recipients has been observed among burnout victims. Since emotional control is decreased, bursts of anger may occur. Finally, the professional does not feel comfortable at work and, not surprisingly, job satisfaction is low.

2. Cognitive symptoms

First and foremost, the burned-out professional feels helpless, hopeless and powerless. Sometimes they even have the fear of 'going crazy' because they feel out of control. Work loses its meaning and after being unsuccessful in changing their work situation, the individual now feels 'trapped' – there is nowhere left to go. A sense of failure is experienced, as well as a feeling of insufficiency and impotence, which may lead to poor job-related self-esteem ('I'm no good for this job'). The professional is preoccupied with him- or herself and feels guilty because (s)he is not able to perform as usual on the job. At this point, suicidal ideas may develop, though this is rarely observed.

In addition, particular cognitive skills might be impaired. For instance, the burned-out professional may not be able to concentrate for a long period. (S)he is forgetful and makes all kinds of minor mistakes and errors in letters, files, notes, meetings, and interviews. Appointments are forgotten and complex and/or multiple tasks are difficult to perform, for example making a phone call and simultaneously looking for a file in a drawer. Thinking becomes more rigid, schematic and detached: black and white opposites are contrasted instead of making distinctions between various shades of grey. Emotional and personal issues and problems are intellectualised. Decision making becomes increasingly difficult, for instance, it becomes difficult for a psychotherapist with burnout to decide when to conclude a therapy. Instead of actively solving problems there is a tendency to run away from reality by daydreaming and fantasising. As a result, loneliness may develop. The individual's frustration tolerance is diminished which increases the likelihood of aggressive behaviours.

One of the most characteristic symptoms of burnout at the interpersonal level is the decreased involvement with recipients. Cognitively, this is reflected by a cynical and dehumanising perception of recipients characterised by negativism, pessimism, lessened empathy and stereotyping. For instance, the nurse who treats her patients as impersonal objects – 'That ulcer from room 34'; the physician who doesn't take seriously the problems of his patients 'who all want to be treated for trifles'; or the prison guard who states that 'these buggers just get what they deserve'. Such negative beliefs are particularly striking since initially the relationships with recipients have been characterised by involvement, concern, and understanding. Instead, the burned-out professional labels his or her recipients in derogatory ways and engages in stereotyping. They might even blame their recipients for their own fate. For instance, the police officer who blames a woman who has been raped for going out alone at night. By derogating, stereotyping and blaming recipients a psychological distance is created which protects or enhances the self. This might foster an air of grandiosity – one feels far above the stupid pupils, the miserable clients, or the pitiable patients – or an air of righteousness ('I've said one hundred times before that the program wouldn't work'). But another pattern is also observed: professionals who believe that they have to deal with problems at work that cannot be solved may come to feel like a martyr ('I was going to be the Martin Luther King of the drunks, leading them to the promised land'). Furthermore, the burned-out professional is resentful and shows hostility and suspicion, not only towards recipients, but also towards colleagues and supervisors. The professional's frustrations, anxieties and problems are projected to others and paranoia might even develop. The professional who suffers from burnout is likely to experience a lack of personal effectiveness which fosters his or her cynicism about the work-role ('You never get anything done right in this damned bureaucracy' or 'As social worker one has to clean society's mess'). Burned-out professionals feel neither appreciated by the organisation nor by their colleagues. They lose their concern for the organisation and become hyper-critical – distrusting management, peers and supervisors.

3. Physical symptoms

Physical symptoms of burnout can be grouped into three categories. First, all kinds of indefinite physical distress complaints are observed like headaches, nausea, dizziness, restlessness, nervous tics, and muscle pains, particularly neck and lower back pain. In some cases hyperventilation occurs. Hyperventilation in its turn may cause peculiar sensations including prickling limbs, dry throat, heart palpations, and heavy perspiration. Generally, hyperventilation is accompanied by anxiety since the individual is afraid of losing control over his or her body. In addition, sexual problems, sleep disturbances, sudden loss or gains of weight, and shortness of breath are reported by those who suffer from burnout. It is claimed that burned-out women often have problems with their menstrual cycle. Chronic fatigue is the most common physical sign of burnout and it is mentioned by virtually every author who describes the syndrome. Those who are burned-out feel extremely tired and physically exhausted.

Their drowsiness and bodily weakness remain all day from the moment they get up in the morning. As we will see in Chapter 3, many of these physical complaints have been identified as symptoms of neurasthenia.

In the second category we find psychosomatic disorders like ulcers, gastric-intestinal disorders, and coronary heart disease. Less serious, but more frequently occurring are prolonged colds and flu that cannot be shaken off. An increased suscept-ibility to such viral infections is considered to be a consequence of prolonged stress. Stress impairs the immune system because it reduces the level of particular hormones in the blood (e.g. ACTH and cortisol) which are essential for keeping the body's immunological defence mechanisms intact. In a somewhat similar vein, flare-ups of pre-existing disorders like asthma, diabetes or rheumatoid arthritis are sometimes observed. Because many burned-out individuals tend to be engaged in risk-taking behaviours, injuries such as fractures (downhill skiing, etc.) are likely to result.

Finally, a number of physiological reactions have been associated with burnout. These include increased heart rate and respiration rate, hypertension, high levels of serum cholesterol, and a decrease of the electrical resistance of the skin due to increased perspiration (the electricdermal response). These are typical physiological stress reactions and they have been demonstrated in numerous laboratory experi-ments with healthy non-burned-out humans, as well as with animals.

4. Behavioural symptoms

Behavioural symptoms are mainly caused by the individual's increased level of arousal. The burned-out professional is hyperactively running around the place, not knowing where to go or what to do, and not being able to concentrate on anything in particular. (S)he acts directly and impulsively without carefully considering altern-ative options. On the other hand, the opposite – procrastination, doubt and indecis-iveness – is also observed. Furthermore, the consumption of stimulants like coffee and tobacco increases, as may the use of alcohol, tranquillisers, barbiturates or illicit drugs. These substances are used in an attempt to reduce job-related tension. However, this strategy is not very effective because it impairs the professional's health and may lead to substance abuse or addiction. Thus, in creating even more tension, a vicious circle might develop. Over- and under-eating are also mentioned as behavioural symptoms of burnout. Much like substance abuse, overeating results from diminished impulse control. Accident-proneness is reinforced by engaging in high-risk-taking behaviours (e.g. fast driving). It is also claimed that those who are burned-out particularly like hazardous pastimes (e.g. scuba diving, sky diving). This makes them experience the type of thrill they miss so deeply at their jobs. By contrast, others abandon their recreational activities because they are over-involved in their jobs, because they feel too exhausted, or because they want to be left alone. Finally, it is observed that burnout victims are often compulsive complainers, although the opposite – denial – is documented perhaps even more frequently.

At the interpersonal level, two patterns emerge. First, a tendency exists towards aggressive or even violent behaviour because the professional's impulse control is weakened. Sudden and unexpected outbursts of rage may occur: doors are being slammed, voices are raised, and other people are shouted at. It is therefore not surprising that interpersonal conflicts may develop, both at work and off work. Often the individual's aggressiveness comes as a real surprise ('I didn't think of myself as one who might react in such a terrible way'). In the helping professions in particular, aggressive behaviours are considered inappropriate and unprofessional.

Second, and much more common, social isolation and withdrawal occur. Typically, the burned-out professional withdraws physically as well as mentally from social contacts at work and is therefore in danger of isolating him- or herself. For instance, the nurse who spends more time alone in the office than together with the patients. One of the most obvious characteristics of burnout is the decreased involvement with recipients. The initial zest and vigour has turned into its opposite: the professional now responds in a detached and mechanical manner. For instance, among police officers the so-called 'John Wayne syndrome' is observed: playing the tough guy who is not moved or touched by anything he gets involved with during his duty. Also, psychological distancing devices are used such as jargon or sick humour, for example the type of humour that the surgeons use in the popular war comedy M*A*S*H. Occasionally, the professional's personal attitude of hopelessness, helplessness and meaninglessness is communicated verbally and/or nonverbally. Needless to say, this strains the helping relationship considerably.

In addition, relationships with colleagues and supervisors are disturbed and conflicts develop that cannot be solved easily. Dealing with burned-out co-workers is difficult since it requires much patience and diplomacy. Burned-out professionals are a heavy burden for the organisation and its members to carry. Not only do they cause extra work, but they also have a negative and demoralising impact on the ambience because of their persistent pessimism and cynical attitudes. Accordingly, a process of symptom contagion might develop and burnout symptoms might be taken on by others at work (see Chapter 5). Nevertheless, their negative behaviour doesn't make them very popular, to say the least. It is therefore not surprising that burned-out professionals are often ignored by their colleagues and supervisors. This strengthens the 'victim's' mistrust and suspicion which fosters social isolation and encourages interpersonal conflicts.

Behavioural burnout-symptoms can also be found off the job. Burned-out professionals might take their work problems home. These problems come to dominate family life and increase interpersonal conflicts with spouse and children. Others, including the partner, who have more attractive and rewarding jobs are envied and thus jealousy develops. As a consequence, the quality of the interpersonal relations is negatively affected. Burned-out professionals feel left behind. They believe that nobody understands what they are going through. In an attempt to avoid negative spill-over effects, work and private life are strictly separated (compartmentalisation). However, this strategy might increase social isolation because, as a result, colleagues are not met informally outside the workplace.

At the organisational level, burnout is first and foremost characterised by reduced effectiveness, poor work performance, and minimal productivity. For instance, fewer clients are being helped and it takes longer to finish a particular case. However, not only does the quantity of the performance deteriorate, so does its quality. More mistakes are made and the work is done less accurately: e.g. patients receive the wrong medication and files are not kept properly. Withdrawal behaviours like lateness, turnover, increased sick-leave and absenteeism indicate the professional's poor commitment. Possibly, these behaviours result from feelings of inequity and resentment ('I work my ass off, but I'm being thwarted by the organisation'). By coming in too late to work, leaving too early, staying home pretending to be sick, or leaving the organisation, the equity balance with the organisation is restored (see Chapter 5). Similarly, theft by the employee can be viewed from the perspective of restoring equity. Frequent clock watching, 'going by the book', being over-dependent on one's supervisors, scepticism ('the house cynic'), and an exceptionally strong resistance to change all signify withdrawal and poor commitment. As a result of exhaustion, poor decision making, and indifference, more accidents may occur. Finally, it is claimed that particular skills are impaired like the ability to organise and the ability to manage one's time adequately. These various negative changes in organisational behaviour are all the more remarkable since initially the professional performed well and had been quite successful in his or her job.

5. Motivational symptoms

Typically, the professional's intrinsic motivation has vanished: zeal, enthusiasm, interest, and idealism are lost. By contrast, disillusionment, disappointment and resignation set in. In nursing the so-called 'Florence Nightingale syndrome' is observed – the romantic image of the profession does not match up to reality. Instead of saving lives heroically like 'the lady with the lamp' once did on the battle fields of the Crimean War, the nurse finds herself performing dull and routine tasks. Since the initial expectations did not come out right, the professional is overwhelmed by disappointment and feels demoralised.

At the interpersonal level this deeply rooted motivational crisis is expressed by a loss of genuine interest in recipients, indifference, and discouragement. The burned-out professional is 'sick and tired' of all those recipients who ask for help, support, advice, attention, or care. Consider teachers who believe that they are merely 'casting pearls before swine', or social workers who strongly believe that they 'have to deal with hopeless cases that cannot be helped anyway'. Occasionally, recipients are scrupulously misused to meet personal and social demands. The professional's present poor and/or inappropriate motivation stands in sharp contrast to his or her initial idealism and drive. Over-involvement with recipients is sometimes observed as an early stage of burnout, which is followed by disillusionment and discouragement.

The professional used to be a loyal, committed and dedicated member of the organisation who showed initiative and who worked hard and efficiently. Unfortunately, this has changed dramatically and the reverse seems to be true now: work motivation is poor, there is a strong resistance to go to work and the professional's initiative is dampened. Moreover, enthusiasm and involvement have made way for resignation, withdrawal, and low morale.

Some critical notes

This enumeration of 132 possible burnout symptoms illustrates that such a laundry list approach is utterly inappropriate to defining the phenomenon. Yet, it shows how the concept is expanded to the extent that it covers virtually anything and therefore loses its meaning. After scrutinising Table 2.1 the reader might find it difficult to come up with any symptom that is not included! Therefore, it seems appropriate to make some critical notes in order to place the multitude of burnout symptoms into perspective.

First, as indicated before, most symptoms result from uncontrolled observations rather than from careful empirical study. In this way, we cannot be sure about the validity of all symptoms that are listed.

Second, many possible symptoms are rather indefinite. For instance, similar mental, physical and behavioural distress symptoms can be observed after such stressful life-events as unemployment or death of a spouse. This non-specific nature of some burnout symptoms is also illustrated by the fact that occasionally opposite symptoms are listed (e.g. over- and under-eating, or displaying an air of grandiosity and feeling helpless). This exemplifies the existence of individual differences in response to stress. Some individuals, when under stress, empty out their refrigerator whereas others cannot swallow down anything at all.

Third, Table 2.1 includes symptoms that characterise different stages of burnout, which might explain some contradictions. It is, for instance, entirely possible that the burnout process starts with over-involvement, which at a later stage makes way for its opposite – underinvolvement or lack of commitment. Thus, opposite symptoms like over- and under-involvement might reflect different phases in the burnout process. We will return to this issue in the next section.

Fourth, it is likely that various patterns are concealed in the myriad symptoms. For instance, it has been suggested that two types of burnout exist: an active and a passive type (Gillespie, 1981). The former is characterised by a high level of arousal, aggressiveness, anger, irritability, and hyperactivity, whereas the latter is characterised by a low level of arousal, social isolation and withdrawal, boredom, and procrastination.

Last but not least, Table 2.1 illustrates the confusion of symptoms, precursors, and consequences of burnout. In the pioneering phase of research into burnout, when clinical observations prevailed and on which Table 2.1 is largely based, this confusion was most obvious. Later, when standardised instruments were used to assess burnout, it became more easy to distinguish between symptoms, precursors, and consequences. However, basically, this distinction remains rather arbitrary since it depends on the definition of burnout that underlies the particular instrument that is used. For instance, social isolation can be considered as a symptom of burnout (as in the case of Miss Jones – see Box 1.2), or as a consequence (as in the case of Mr Dijkstra – see Box 2.1), or even as a precursor (as in Cherniss' professional mystique of collegiality). Making a distinction between symptoms, precursors and consequences of burnout boils down to drawing an arbitrary line somewhere on a continuous scale. Ultimately, this is a matter of definition as we will see in the next section.

2.2 DEFINITIONS: STATE AND PROCESS

In most early writings burnout was 'defined' by merely summing up its symptoms (see Box 1.3). Such laundry lists are problematic since they are inevitably selective. Clearly, it is quite impossible to include all symptoms of burnout into one definition! In addition, by listing symptoms a rather static picture emerges of burnout as a particular negative mental state instead of a process that develops over time. In principle, the first drawback can be avoided by selecting the most characteristic core symptoms of burnout as is done in so called state definitions. The second drawback can be avoided by describing the dynamic process of the burnout syndrome, as is done in so called process definitions. In the following sections, the most important examples of both types of definitions are presented. Of course, both types of definitions are not mutually exclusive. Rather, they are complementary in the sense that state definitions describe the end-state of the burnout process. In Chapter 5 where we discuss theoretical approaches to burnout, we will return in greater detail to the issues raised by the following definitions.

State definitions

The two most common state definitions of burnout are presented below, supplemented by a less widely used, but somewhat more specific description of the syndrome. Together, these three definitions describe the most crucial elements of burnout as a negative mental condition.

Burnout as a multidimensional syndrome

Probably the most often cited definition of burnout comes from Maslach and Jackson (1986, p. 1):

> Burnout is a syndrome of emotional exhaustion, depersonalisation, and reduced personal accomplishment that can occur among individuals who do 'people work' of some kind.

Its popularity is due to the fact that the most widely used and well validated self-report questionnaire, the Maslach Burnout Inventory (MBI), includes the three dimensions of this definition (see Chapter 3). **Emotional exhaustion** refers to the depletion or draining of emotional resources. Professionals feel that they are no longer able to 'give' themselves at a psychological level – they feel at the end of the rope. **Depersonalisation** points to the development of negative, callous, and cynical attitudes toward the recipients of one's services. They are labelled in derogatory ways and treated accordingly. The term depersonalisation might cause some confusion since in psychiatry it is used to denote a person's extreme alienation from the self and from the world. For instance, the person observes his or her own actions like a spectator. However, in Maslach and Jackson's definition, depersonalisation refers to an impersonal and dehumanised perception of *recipients*, rather than to an impersonal view of the *self*. Therefore, dehumanisation would have been a more appropriate term. It was probably not chosen because of its rather extreme connotations. Finally, **lack of personal accomplishment** is the tendency to evaluate one's work with recipients negatively. It is believed that the objectives are not achieved, which is accompanied by feelings of insufficiency and poor professional self-esteem. Most importantly, Maslach and Jackson (1981b; 1986) initially claimed that burnout exclusively occurs in occupational groups where professionals deal directly with recipients, like students, pupils, clients, patients, customers, or delinquents. Hence, from their point of view burnout is mainly restricted to the human services like health care, education, and social work. However, Maslach and her colleagues meanwhile expanded the burnout concept beyond the human services (Maslach and Leiter, 1997). Burnout is redefined as a crisis in one's relationship with work, not necessarily as a crisis in one's relationship with people at work. It includes three somewhat more general aspects that do not explicitly refer to 'people work': exhaustion, cynicism, and professional efficacy. Recently, an adapted version of the MBI was developed to measure burnout outside the human services – the MBI-General Survey (Schaufeli *et al.*, 1996a).

Burnout as exhaustion

Pines and Aronson (1988) present a slightly broader definition of burnout. They include physical symptoms as well and in their view burnout is not restricted to the human services. They describe burnout as 'a state of physical, emotional, and mental exhaustion caused by long-term involvement in situations that are emotionally demanding' (p. 9). **Physical exhaustion** is characterised by low energy, chronic fatigue, weakness, and a wide variety of physical and psychosomatic complaints. **Emotional exhaustion** involves feelings of helplessness, hopelessness, and entrapment, which in extreme cases can lead to emotional breakdown. Finally, **mental exhaustion** refers to the development of negative attitudes towards one's self, one's work, and life itself. Since excessive emotional demands are not restricted to the human services, burnout is also observed in other occupational settings (e.g. management) as well as outside work, for example in love and marriage (Pines, 1996) and in political activism (Pines, 1994). Initially, Pines and Aronson (Pines *et al.*, 1981) distinguished between 'burnout' and 'tedium'. Much like Maslach and Jackson (1981b; 1986), the former was defined as the result of continuing *emotional* pressure associated with an intense involvement with *people* in human services professions. The latter was defined more broadly, as the result of *any* prolonged chronic physical, emotional, or mental pressures. Pines and Aronson (Pines *et al.*, 1981) claimed that the symptoms of burnout and tedium are similar, but that they have different causes. However, for reasons unexplained, in the second edition of their 1981 book this distinction has been abandoned (Pines and Aronson, 1988). Pines and her colleagues also developed a brief self-report questionnaire to measure their construct, which was initially labelled 'Tedium Measure' (TM; Pines *et al.*, 1981) and later called 'Burnout Measure' (BM; Pines and Aronson, 1988). As we will see in the next chapter, psychometric studies with the TM/BM did *not* confirm the distinction between the three previously mentioned aspects of exhaustion.

Burnout as a dysphoric, work-related state in normals

Finally, a less well-known but most precise definition of burnout has been proposed by Brill (1984, p. 15):

> Burnout is an expectationally mediated, job-related, dysphoric and dysfunctional state in an individual without major psychopathology who has (1) functioned for a time at adequate performance and affective levels in the same job situation and who (2) will not recover to previous levels without outside help or environmental rearrangement.

This description of burnout entails four key elements. First, two core symptoms of burnout are specified: dysphoric symptoms and reduced job performance. Second, a clear statement is made about the cause of burnout: unmet expectations play a major role. According to Brill (1984) these expectations may be either related to the job or to oneself. For instance, the job might appear to be less challenging, creative, varied, or stimulating than envisioned. On the other hand, the individual might come to perceive him/herself as being less tolerant, empathic, patient, or understanding than initially expected. Third, the definition specifies the severity or intensity of the burnout symptoms since only individuals are included who cannot cope without outside help. Finally, the definition of Brill includes two criteria – work relatedness and absence of psychopathology – that exclude burnout outside the occupational context and burnout among individuals who suffer from mental illness (e.g. depression), respectively. By narrowing the definition of burnout, not only in terms of symptomatology and aetiology, but also in terms of excluding particular contexts and groups, Brill (1984) made a most important contribution to the specification of the burnout-syndrome.

Conclusion

State definitions of burnout differ in scope, precision, and dimensionality of the syndrome. Taken together, three elements seem to be crucial that refer to the symptomatology, aetiology and the domain of burnout, respectively.

- Dysphoric symptoms – most notably emotional or mental exhaustion – lie at the core of the syndrome. Moreover, negative attitudes towards others, and decreased effectiveness and performance are observed.

- Inappropriate expectations and emotional demands play a major role in the development of burnout.

- Burnout is generally considered to be work-related and it occurs in 'normal' individuals who do not suffer from psychopathology and who have functioned at adequate levels before.

Process definitions

Four process definitions of burnout are discussed below. Each of these definitions emphasises a particular aspect of the burnout process.

Burnout as a three-stage process

Almost two decades ago, Cherniss (1980a, p. 5) was among the first who proposed a straightforward description of the burnout process:

> Burnout refers to a process in which the professional's attitudes and behaviour change in negative ways in response to job strain.

In another publication that appeared in the same year, Cherniss (1980b, p. 17) is somewhat more specific as he describes burnout as a three-stage process:

> The first stage involves an imbalance between resources and demands (stress). The second stage is the immediate, short-term emotional tension, fatigue, and exhaustion (strain). The third stage consists of a number of changes in attitude and behaviour, such as a tendency to treat clients in a detached and mechanical fashion, or a cynical preoccupation with gratification of one's own needs (defensive coping).

Since the nature of the job demands are not specified, burnout can basically occur in each and every occupation, although in his own empirical work Cherniss (1980a; b; 1995) deals exclusively with the human services professionals. An important new aetiologic element is introduced by Cherniss in the third stage of burnout: the individual's way of coping with stress. Although Cherniss considers excessive job demands as the root cause of burnout, a defensive coping strategy, which is charac- terised by avoidance and withdrawal, fosters its development.

Burnout as increasing disillusionment

Edelwich and Brodsky (1980) described burnout as a process of increasing disillu- sionment. They define burnout in the helping professions as 'progressive loss of idealism, energy, and purpose experienced by people in the helping professions as a result of conditions in their work' (p. 14). The authors consider the helpers' initial idealistic expectations and noble aspirations as built-in sources of future frustration and therefore as major causes of burnout. In Chapter 5, their four-stage process- model is discussed in greater detail.

Burnout as psychological erosion

Another important element is emphasised by Etzion (1987). Unlike most other authors, she maintains that minor sources of stress are crucial to the understanding of burnout. She writes:

In burnout, I propose that continuous, barely recognisable, and for the most part denied misfits between personal and environmental characteristics are the source of a slow and hidden process of psychological erosion. Unlike other stressful phenomena, the mini-stressors of misfit do not cause alarm and are rarely subject to any coping efforts. Thus the process of erosion can go on for a long time without being detected.   (pp. 16–17)

Accordingly, burnout is a slowly developing process that starts without warning and evolves almost unrecognised up to a particular point. Suddenly and unexpectedly, one feels totally exhausted while being unable to relate this devastating experience to any particular stressful event. Etzion (1987) argues that it is this treacherous nature that makes burnout so special. Recently, and similarly, burnout was charac-terised as 'the erosion of the soul' by Maslach and Leiter (1997). They write:

It represents an erosion in values, dignity, spirit, and will – an erosion of the human soul. It is a malady that spreads gradually and continuously over time, putting people into a downward spiral from which it's hard to recover.   (p. 17)

The process of burning out

Hallsten (1993) presented a complex framework for the process of burning-out. He defines burnout '. . . as a form of depression that results from the process of burning out, which is a necessary cause of burnout' (p. 99). Accordingly, the aetiology, the process of burning out that develops in several phases, and not the outcome (i.e. a particular depressive state) is considered to be specific for burnout. Hallsten assumes that the process of burning out appears '. . . when the enactment of an active, self-definitional role is threatened or disrupted with no alternative role at hand' (p. 99). Or more specifically: 'burning out refers to the recurrent, interactive pattern of anxious-depressive reactions and rigid, pseudoactive strivings' (p. 100). According to Hallsten this pattern which may eventually result in exhaustion, fatigue and depress-ive episodes constitutes the unique component of burnout. However, depending on other aetiologic factors such as personal vulnerability, strength of goal orientation, and perceived environmental congruency, other outcomes of the process may also result. These outcomes may be either positive (i.e. balanced commitment) or negative (i.e. frustration, alienation). It is important to note that in Hallsten's view the process of burning out may occur in any occupational or non-occupational context.

Conclusion

As far as process definitions of burnout are concerned, three conclusions can be drawn.

- Most process definitions of burnout maintain that burnout starts with tensions that result from the discrepancy between the individual's expectations, inten-tions, strivings, and ideals and the demands of the harsh reality of everyday life.

- The stresses that results from such an imbalance develop gradually and they may be consciously experienced by the individual or they may remain unnoticed for a long time.

- The way in which the individual copes with these stresses is crucial for the development of burnout.

Towards a working definition of burnout

It is impossible to present a general description of burnout that agrees with all state and process definitions that were presented above. The reason why such a consensus definition cannot be formulated is evident: the existing definitions are contradictory, at least at some points. For instance, some definitions state that burnout exclusively occurs in the human services, whereas other less restrictive definitions assume that burnout can be found in other occupations as well, or even outside the occupational field. Despite such inconsistencies, we feel that it is important to formulate a working definition of burnout that will guide us through this book. Our definition is based on the review of burnout symptoms, as well as on the previous state and process definitions:

> Burnout is a persistent, negative, work-related state of mind in 'normal' individuals that is primarily characterised by exhaustion, which is accompanied by distress, a sense of reduced effectiveness, decreased motivation, and the development of dysfunctional attitudes and behaviours at work. This psychological condition develops gradually but may remain unnoticed for a long time by the individual involved. It results from a misfit between intentions and reality in the job. Often burnout is self-perpetuating because of inadequate coping strategies that are associated with the syndrome.

This working definition of burnout specifies its general symptomatology, its preconditions, as well as the domain in which it occurs. More specifically, the definition narrows down over 100 burnout symptoms to one core indicator (exhaustion) and four accompanying, general symptoms: (1) distress (affective, cognitive, physical, and behavioural); (2) a sense of reduced effectiveness; (3) decreased motivation; (4) dysfunctional attitudes and behaviours at work. Furthermore, frustrated intentions and inadequate coping strategies play a role as preconditions in the development of burnout and the burnout process is considered to be self-perpetuating despite the fact that it may not be recognised initially. Finally, the domain is specified: the symptoms are work-related and burnout occurs in 'normal' individuals who do not suffer from psychopathology.

2.3 OLD WINE IN NEW BOTTLES?

Burnout has been equated with a host of alternative terms (in alphabetical order): alienation, anxiety, boredom, chronic fatigue, depression, effort syndrome, 'flame-out', 'giving-up-given-up' complex, hyperaesthetic-emotional syndrome, job dissatisfaction, job stress, 'nerves', nervous breakdown, neurasthenia, neurocirculatory asthenia, over-strain, surmenage, tedium, vital exhaustion, and 'worn-out' (Maslach and Schaufeli, 1993). Obviously, the distinctiveness of burnout from such psychological constructs and popular lay-terms is an important issue. In other words, we have to answer the question 'Is burnout a truly "new" phenomenon or is it merely old wine in new bottles?' Since it would go beyond the scope of this chapter to answer this question for all notions mentioned above, we concentrate on the three most important ones: job stress, depression and chronic fatigue.

In discussing the distinction between burnout and these three concepts, two prob-lems emerge. First, these concepts are plagued by the same sort of definitional ambiguity as is burnout. For instance, Cox (1978, p. 1) writes:

> The concept of stress is elusive because it is poorly defined. There is no single agreed definition in existence. It is a concept which is familiar to both layman and professional alike; it is understood by all when used in a general context but by very few when a more precise account is required, and this seems to be the central problem.

One could have used the same words to describe the conceptual confusion that is surrounding burnout! Second, it is clear that burnout can only be distinguished from other related concepts in a relative way. Trying to establish sharp boundaries would be artificial and therefore unwise. However, at what point does a relationship become so strong that both concepts have to be considered as indices of a similar underlying factor? Although this is a legitimate question to ask, a clear answer cannot be given because ultimately it is matter of judgement and definition that depends on one's theoretical assumptions.

Burnout and job stress

Burnout can be considered as a particular kind of prolonged job stress. An indi-vidual is experiencing job stress when the demands at the workplace tax or exceed his or her adaptive resources (*cf.* Lazarus and Launier, 1978, p. 296). The longer time perspective that is introduced to make the distinction between stress and burn-out is also implied in the terminology: burning out by depleting one's resources is by definition a long-term process.

According to Brill (1984), stress refers to a temporary adaptation process that is accompanied by mental and physical symptoms. This is illustrated by A and B in Figure 2.1. Case A represents an individual who experienced stress and returned to his or her normal level of functioning again. Case B is on the right track, but has not reached his or her initial level yet. Burnout, on the other hand, refers to a breakdown in adaptation accompanied by chronic malfunctioning at work. This is illustrated by cases C and D. C represents an individual who collapsed and is malfunctioning on a stable level, whereas D is still in the process of deteriorating. Accordingly, a relative distinction can be made between stress and burnout when the process is taken into account as it develops over time. This suggests that both concepts can exclusively be discriminated retrospectively when the adaptation has

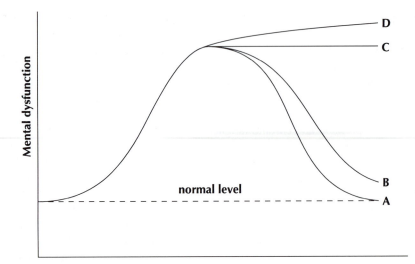

Figure 2.1   Stress (A, B) and Burnout (C, D) (adapted from Brill (1984, p. 21))

been successfully performed (job stress) or when a breakdown in adaptation has occurred (burnout). Thus, burnout and job stress can be distinguished retrospectively when the time frame is taken into account.

Yet another more basic distinction between burnout and job stress can be made when the former is defined as a multidimensional syndrome that includes, in addition to energy depletion, the development of negative, dysfunctional attitudes and behaviours at work. It is generally acknowledged that stress responses at work include physical, psychological (affective, cognitive, motivational), and behavioural symptoms, similar to those listed in Table 2.1. For instance, in their seminal review on stress in organisations, Kahn and Byosiere (1992) identified about a dozen psychophysiological stress responses, nearly 40 psychological stress responses, and over a dozen behavioural stress responses. Dysfunctional attitudes and behaviours at work that are typical for burnout were not listed, though. It seems therefore that burnout is a particular, multidimensional stress-response that includes characteristic negative, job-related attitudes and behaviours that are not covered by the traditional job stress concept. Or as Cordes and Dougherty (1993) put it: emotional exhaustion is a traditional stress variable; depersonalisation is a new construct not formerly appearing in the stress literature; and reduced personal accomplishment adds the assertion that self-evaluations are central to the stress-experience. This implies that burnout is a unique, multidimensional, chronic stress reaction that goes beyond the experience of mere exhaustion. Schaufeli and Van Dierendonck (1993), showed that emotional exhaustion is indeed related to physical and psychological stress responses in nurses, whereas depersonalisation and reduced personal accomplishment constitute a separate attitudinal component of burnout. Roughly speaking, emotional exhaustion shares about 30% of the variance with these two stress responses, whereas for depersonalisation and reduced personal accomplishment the shared variance is only 14% and 10%, respectively. Percentages of shared variance have to be interpreted as follows: if two concepts were identical they would share 100% of their variance, whereas two completely independent concepts would share 0%. Generally speaking, in social scientific research 10% of shared variance is considered to indicate a modest relationship, whereas 30% indicates a substantial relationship between two concepts.

Finally, it has been claimed that everybody can experience stress, while burnout can only be experienced by those who entered their careers enthusiastically with high goals and expectations and who restlessly pursue success in their jobs. Initially, burnout candidates work hard and are strongly involved and committed to their job. An individual who has no such initial motivation can experience job stress, but no burnout. This concurs with the belief that burnout is expectancy mediated, as we have seen previously in the definition of Brill (1984). More generally, Pines (1993) argues that individuals who expect to derive a sense of significance from their work are susceptible to burnout. Those with no expectations of that kind would experience job stress instead of burnout. So, once again, job stress and burnout can only be discriminated retrospectively.

Burnout and depression

From the outset, the relationship between burnout and depression has been debated. For instance, Freudenberger (Freudenberger and Richelson, 1990) claimed that (reactive) depression is most often accompanied by guilt, whereas burnout generally occurs in the context of anger. Unfortunately, he presents only clinical evidence for this assertion. Moreover, Freudenberger argues that the symptoms of burnout, at least initially, tend to be job-related and situation-specific rather than pervasive. In the early stages of burnout, people often still feel happy and are productive in other areas. Instead, according to Freudenberger, 'real' depression is characterised by a generalisation of the person's symptoms across *all* situations and spheres of life. In a somewhat similar vein, Warr (1987) has distinguished between two types of affective well-being: depression is considered to be 'context-free', whereas burnout is regarded as 'job-related'. Oswin (1978) described a syndrome of 'professional depression' among nurses which bears close resemblance to burnout, including being over-tired, becoming hardened, and accepting one's ineffectuality at the job. Thus, on a conceptual level there seems to be some agreement about the specificity of burnout as a job-related syndrome, which is characterised by dysphoric symptoms that are similar to those of depression.

Empirically speaking, research on the discriminant validity of both concepts shows that particularly the emotional exhaustion component of burnout is related to depression: both share about 25% of their variance (see also Chapter 4). The relationships with other burnout components like depersonalisation and personal accomplishment are less strong: they share less than 10% of their variance. Recently, Glass and McKnight (1996) reviewed 18 studies of burnout and depression that included over 4800 subjects and they concluded:

Burnout and depressive symptomatology are not simply two terms for the same dysphoric state. They do, indeed, share appreciable variance, especially when the emotional exhaustion component is involved, but the results do not indicate complete isomorphism. We conclude, therefore, that burnout and depressive symptomatology are not redundant concepts. (p. 33)

It is unlikely that the relatively strong association between emotional exhaustion and depressive symptomatology is due to overlap in item content of the scales used since factor-analytic studies show that different burnout and depression factors emerge when the items of the burnout and depression instruments are pooled (Leiter and Durup, 1994; McKnight, 1993). Finally, a study among nurses suggested that burnout leads to depression instead of the other way around (Glass *et al.*, 1993).

Burnout and chronic fatigue

Since the 1980s there has been a growing interest in the so-called Chronic Fatigue Syndrome (CFS), although earlier descriptions date back to the 1930s. The renewed interest for this apparently widespread phenomenon has led some to call it somewhat derogatorily 'yuppie flu'. The most prominent symptom of CFS is persistent unexplained fatigue, whereas other symptoms commonly reported include: mild fever or chills, sore throats, painful lymph nodes, unexplained generalised muscle weakness, muscle discomfort, prolonged generalised fatigue after levels of exercise, generalised headaches, joint pain, neuro-psychiatric complaints, and sleep disturbances (Jason *et al.*, 1995). Recently, a joint working group of the (British) Royal Colleges of Physicians, Psychiatrists and General Practitioners (1996) stated in their excellent report that CFS is characterised by:

> . . . a minimum of 6 months of severe physical and mental fatigue and fatiguability, made worse by minor exertion. Other symptoms such as muscle pain, sleep disorder and mood disturbance are common. (p. 41)

Unlike burnout, CFS is pervasive. It can affect virtually all major bodily symptoms: neurological, immunological, hormonal, gastrointestinal and musculoskeletal problems have been reported (Jason *et al.*, 1995). Burnout symptoms are primarily psychological, although accompanying physical symptoms are not uncommon. By contrast, CFS primarily includes physical symptoms, although accompanying psychological symptoms are observed as well. Therefore, some authors propose to use physical criteria (i.e. low-grade fever, nonexudative pharyngitis, tender cervical or axillary lymph nodes) in addition to debilitating fatigue to diagnose CFS (Jenkins, 1991). However, others maintain that there are no objective abnormal physical signs that can be used diagnostically (Royal Colleges, 1996). Although the debate about the proper diagnostic criteria continues, it is clear that physical symptoms are much more prominent in CFS than in burnout. Furthermore, burnout is job-related, whereas CFS is not restricted to a particular life sphere. Generally, burned-out workers blame their jobs for the condition they are in. In their experience a clear link exists between their current mental state and their job. For patients who suffer from CFS, the origin of their symptoms is unclear. In fact, severe unexplained fatigue and exhaustion is a hallmark of CFS. That is what makes CFS such a bewildering experience. Finally, while exhaustion is a common denominator of CFS and burnout, the development of negative, dysfunctional attitudes and behaviours characterise the latter but not the former.

To date, no empirical studies exist on the relationship between burnout and CFS. However, it appears that approximately half of the CFS patients fulfil criteria of affective disorder, even when the symptom of fatigue is removed from the criteria of mood disorder (Royal Colleges, 1996). Thus, depression and CFS seem to overlap to a considerable extent.

Conclusion

Is talking about burnout not just putting 'old wine in new bottles'? We do not believe so. Burnout can be distinguished from stress in several ways: relatively, with respect to the longer time frame involved, as well as more principally, with respect to its symptomatology and its high initial levels of motivation. In addition, burnout can be distinguished conceptually and empirically from depression and from Chronic Fatigue Syndrome. Not only because burnout is typically job related, and thus less pervasive than depression or CFS, but also because burnout includes social and attitudinal symptoms that are absent in both of the other syndromes. Moreover, CFS is characterised much more by physical symptoms. Therefore, it seems that burnout is a genuine phenomenon, which requires serious attention from researchers, practitioners and administrators. In Chapter 3 we will return to the issue of specificity of burnout when we discuss the assessment of the syndrome.

2.4 SUMMARY

Over 130 possible burnout symptoms are identified. These symptoms, which, for the most part, result from clinical impressions, are grouped according to their nature (i.e. affective, cognitive, physical, behavioural, and motivational) as well as according to their level (i.e. individual, inter-personal, and organisational). However, despite this systematic classification, merely listing all possible burnout symptoms is not an adequate way to define the syndrome. It mainly illustrates the almost all-inclusiveness of the notion and it further emphasises the need for an operational definition. The most prominent state definitions and process definitions of burnout share some common elements that are included in a working definition of burnout that stresses its multidimensional nature as well as its work-relatedness. As far as the process of burnout is concerned, this working definition highlights the role of the mismatch of intentions and reality at the job and inadequate coping strategies. Although burnout shares indefinite distress symptoms with job stress, dysphoric symptomatology with depression, and severe fatigue with CFS, it can be conceptually distinguished from these three alternative negative conditions. In addition, empirical evidence strongly suggests that burnout and depression in particular are distinct, albeit related constructs.

CHAPTER THREE

How and where to find it?

Assessment and prevalence

How can burnout be assessed? What methods and instruments can be used to measure it? And how valid and reliable are these techniques? The first aim of this chapter is to present a comprehensive and critical overview of the way burnout may be assessed. We will use our working definition of burnout from the previous chapter as a provisional yardstick in order to see what core elements of burnout are covered by the proposed instruments. Since most empirical research either employs the Maslach Burnout Inventory (MBI) or the Burnout Measure (BM), these questionnaires are discussed in somewhat greater detail. The second aim of this chapter is to address the issue of the prevalence of burnout. To what extent can the occurrence of burnout be estimated? Also, do particular burnout profiles exist among different occupational fields or professions?

3.1 ASSESSMENT TOOLS: A BRIEF INTRODUCTION

Basically, psychological characteristics can be assessed by observation, interview or self-report. In addition, physiological parameters such as heart rate, blood pressure, and adrenaline or cortisol levels can be employed. Although few studies use such parameters (see Chapter 4) no specific psychophysiological, psychoneuroendocrine or psychoimmunological markers for burnout have yet been identified that can be used for assessment purposes. Therefore, physiological parameters will be excluded from further discussion.

Observation

Observation is the most obvious way to gather information about an individual's psychological characteristics. As we saw in Chapter 1, burnout research started with the observations of Freudenberger and others. However, their observations were neither systematic nor standardised. For reasons of reliability and validity, individuals should be observed systematically in standard situations using a limited set of specific behavioural criteria. As the assessment centre technique in personnel selection shows, this is a rather tiresome, complicated, and expensive thing to accomplish. Applied to burnout, clear behavioural criteria should be established and operationalised in rating scales, a particular standardised situation must be created that provokes such behaviour, and assessors have to be trained in observing these behaviours using the rating scales. To date, no such standardised behavioural observation protocols have been developed to assess burnout.

Interview

Interviewing clients or patients in order to assess their mental state is most popular, particularly among physicians, counsellors, and social workers. However, interviews are notoriously unreliable for individual assessment purposes, unless they are well structured. As a research tool, interviews are not only labour intensive, and thus rather inefficient, but also inevitably subjective. This subjective nature is also a strength, though. The interviewer is flexible and may ask the interviewee, for instance, to clarify statements, elaborate particular issues, or explain contradictions. Hence, more in-depth information can be collected by using an interview. In general, the medical profession prefers standardised diagnostic protocols that are based on more or less structured interviews, whereas psychologists traditionally rely heavily on self-report questionnaires.

Self-report

Paper-and-pencil self-report questionnaires have a number of practical benefits, which makes their use very popular. They are quick to administer to large groups and are therefore very efficient and cheap. Moreover, they are easy to administer, to score, and to interpret. In principle, such questionnaires are reliable tools because standardisation eliminates the assessors' subjectivity. However, standardisation has its price: self-reports are inflexible and they are open to answering bias (e.g. a tendency to fake good, or to avoid extreme answers). Besides, research and development takes quite some time and effort, particularly when the questionnaire is meant to be used for individual assessment. Most importantly, however, the validity of self-reports is not beyond question. In other words, one is never entirely sure what psychological characteristic is exactly tapped. The answer has to come from empirical, psychometric research. In order to fully understand the outcomes of such research, we elaborate in the next section on some crucial psychometric notions.

About reliability and validity: some psychometric notions

In discussing the psychometric properties of the burnout instruments we focus on their reliability and validity, respectively. Reliability refers to the accuracy of the measurement: the more reliable an instrument the less the individual's score is influenced by arbitrary and random factors such as the particular wording of an item, or the particular person who administers the instrument.

In the case of self-report questionnaires, the most often reported type of reliability is 'internal consistency'. It refers to the homogeneity of a set of items that constitute a particular (sub)scale. The degree of internal consistency is expressed by Cronbach's α coefficient: high values of α indicate that the set of items is homogenous. The maximum value for α is 1.00. As a rule of thumb, a value greater than .70 is considered sufficient, whereas values exceeding .80 are good. Unfortunately, α depends on the number of items included so that shorter scales are more likely to have lower α-values.

Validity is a generic term that refers to the extent to which the intended construct is actually tapped by the instrument. Or more specifically, does the score on a particular burnout questionnaire actually reflect the individual's level of burnout or does it represent something else? Since there is no single answer to this question several aspects of validity have to be distinguished. Depending on the type of validity, different statistical techniques are used and no general criteria for each of these aspects of validity exist. However, most techniques are ultimately based on the Pearson Product Moment correlation (r).

The correlation coefficient r is a measure of the strength and direction of the association between two variables. It ranges from -1.0 which reflects a perfect negative association of both variables (high values of one variable correspond to low values of the other and vice versa), to $+1.0$ which reflects a perfect positive association (high and low values of one variable correspond to high and low values of the other, respectively). A correlation coefficient of 0.0 reflects a minimal association between two variables: high and low values of one variable do not correspond systematically to high or low values of the other. As a rule of thumb, effects in absolute values of r of about .10 or below are considered to be small (i.e. a weak relationship between two variables), whereas rs of about .30 are considered as medium, and rs of about .50 or greater are considered as large (Cohen, 1977). When the value of r is squared, the amount of shared (explained) variance of the two variables is obtained. Thus, a small correlation corresponds to 1% or less of explained variance, a medium correlation to about 10%, and a large correlation to 25% or more of explained variance. In practice, rs less than .15 (2% of explained variance) are frequently found, whereas a shared variance of psychological variables of more than 30% is already regarded as substantial and rs exceeding .55 are rare.

To answer the question whether one can use a particular instrument to obtain valid measures of the intended construct, several aspects of validity have to be distinguished. For instance, factorial validity refers to the dimensionality of a burnout questionnaire: is the questionnaire able to distinguish between various dimensions or components of burnout? Construct validity refers to the extent to which the intended construct is appropriately measured with regard to other instruments. This can, for instance, be assessed by comparing scores on different questionnaires: correlations between questionnaires intended to measure burnout are expected to be high (convergent validity), whereas correlations of a burnout measure with questionnaires that are intended to measure other constructs (e.g. depression, job satisfaction, work overload) are expected to be low (discriminant validity).

3.2 SELF-REPORT MEASURES OF BURNOUT

A vast array of self-report instruments have been used to assess burnout (cf. Schaufeli *et al.*, 1993a). They differ considerably in scope, use, and the psychometric effort that has been spent on their development.

Do-it-yourself inventories

Because of the popularity of burnout it is not surprising that many do-it-yourself inventories have been published. They can be found in professional journals or in mass media under appealing headings such as: 'How burned-out are you?', 'What's your burnout score?' or 'The burnout-test – Examine your beliefs about work, about leisure, about yourself'. These days do-it-yourself burnout inventories can also be found on the internet.

Perhaps the best known do-it-yourself inventory is the Freudenberger Burnout Inventory (FBI) (Freudenberger and Richelson, 1980, pp. 17–19). It includes items like: 'Do you feel fatigued rather than energetic?', 'Are you increasingly cynical and disenchanted?' and 'Does sex seem like more trouble than it's worth?' The authors also provide an interpretation of the scale scores. For instance, those scoring in the highest category are warned: 'You are in a dangerous place, threatening your physical and mental well-being'.

Typically, do-it-yourself inventories have not been studied empirically at all. Therefore, we are unaware of their psychometric quality. At best, do-it-yourself questionnaires convey a picture of the author's definition of burnout. At worst, the subject becomes alarmed unduly. 'Norms' are entirely arbitrary and not based on empirical research so that the corresponding interpretations are meaningless.

Self-assessment

The following description of burnout includes many aspects of our previously mentioned working definition:

> The tendency for committed physicians to lose enthusiasm for their work and to become less effective in managing the stress of emotional contact with patients. Symptoms may include some of the following – fatigue, withdrawal from patients and colleagues, cynicism, irritability, difficulty relaxing off work, physical manifestations of anxiety and depression, and feelings of diminished enthusiasm and effectiveness at work. (Rafferty *et al.*, 1986, p. 489)

Bearing this description in mind, family practice residents were asked to assess themselves on a nine-point rating scale that ranged from 'not at all burned out' to 'very burned out'. Moreover, each physician completed the MBI. The physician's self-assessment of overall burnout was correlated substantively with the MBI-emotional exhaustion scale ($r = .48$) and somewhat less with the depersonalisation scale ($r = .34$). This suggests that physicians assess burnout predominantly in terms of emotional exhaustion. That conclusion is confirmed by Pick and Leiter (1991) who found elevated emotional exhaustion scores for nurses who assessed themselves 'burned out' compared with those who found that they were 'coping well'. Thus, it seems that in the eyes of the beholder, burnout is principally associated with exhaustion.

Questionnaires with limited application

A number of burnout inventories have been exclusively employed in one or two studies. Three illustrations are presented of an inductively constructed and a deductively constructed multidimensional questionnaire, and of a one-dimensional instrument,[1] respectively.

Blostein *et al.* (1985) analysed about 50 burnout symptoms in child welfare workers and found six underlying dimensions:

1   'classic' burnout (e.g. exhaustion, cynicism);

2   negative feelings about clients (e.g. intolerance, postponement of client contacts);

3   feeling over-stimulated (e.g. inability to relax, problems with sleeping);

4   feeling overwhelmed (e.g. needing to be alone, feeling pulled in all directions);

5   physical problems (e.g. colds and flu, headaches);

6   lack of intimacy (e.g. emptiness, feeling unappreciated).

Typically, this kind of exploratory and inductive approach does not *a priori* specify the dimensions of burnout. Instead, these dimensions are labelled *post hoc* after the data have been analysed. Clearly, three elements from our earlier working definition of burnout can be found: exhaustion (1); distress (3, 4, 5); and dysfunctional attitudes and behaviours (1, 2).

By contrast, the fifteen-item Perceptual Job Burnout Inventory (PJBI) was constructed deductively (Ford *et al.*, 1983). That is, three dimensions of burnout were *a priori* specified:

1   emotional exhaustion and cynicism;

2   demoralised, frustrated feelings and reduced efficiency;

3   excessive demands on energy, strength and resources.

Unfortunately, these dimensions did not emerge in either a social services sample or a corporate sample. Instead, a different pattern of dimensions was found in each sample with only one more or less common dimension that was labelled 'emotional exhaustion' (in the social services sample) and 'fatigue' (in the corporate sample). Accordingly, the factorial validity of the PJBI was not confirmed across samples and only one symptom from our working definition was included: exhaustion/ fatigue.

The Staff Burnout Scale for Health Professionals (SBS-HP) is a one-dimensional measure that includes 30 items, 20 of which assess burnout whereas the remaining ten items form a Lie Scale to detect tendencies to fake good results (Jones, 1980). According to the author, the SBS-HP assesses adverse cognitive, affective, behavioural and psychophysiological reactions that are considered to constitute the burnout syndrome. Hence, a rather limited definition of burnout is used, which excludes, for instance, dysfunctional attitudes and reduced effectiveness.

The internal consistency of the SBS-HP is satisfactory: $\alpha > .80$ (Jones, 1980; Brookings et al., 1985; Beer and Beer, 1992). As expected, SBS-HP-scores are positively related with job stressors (e.g. high patient-to-staff ratios, working undesirable shifts, and receiving little family support) and with behavioural stress reactions (e.g. job turnover, absenteeism, illness, and tardiness), which is indicative for its construct validity. Furthermore, SBS-HP scores are substantively correlated with related constructs (e.g. depression, job dissatisfaction, physical and mental strain, and poor self-esteem), which underscores its poor discriminant validity. Finally, high correlations exist between the SBS-HP score and the MBI-emotional exhaustion ($r = .65$) and the MBI-depersonalisation ($r = .54$) scores, whereas the relationship with the MBI-personal accomplishment scores is moderate ($r = -.33$) (Brookings et al., 1985). These results illustrate the convergent validity of the SBS-HP. However, it should be noted that the findings of the three validity studies are preliminary since all samples are quite small.

Burnout Measure (BM)

Although the test authors define burnout as a three-dimensional construct that comprises physical, emotional and mental exhaustion (see Chapter 2), they conceive of the BM as a one-dimensional questionnaire for which a single composite burnout score is computed (Pines et al., 1981; Pines and Aronson, 1988). Thus, the BM is not a proper operationalisation of their definition of burnout. Since, according to Pines and her colleagues, burnout is not restricted to the occupational domain, slightly adapted versions of the BM were proposed, for instance, to measure 'couple burnout' (Pines, 1996) and even 'national burnout' (Pines, 1994). The latter refers to the public response in Israel among Arabs and orthodox Jews to the Palestinian *Intifada*.

The BM consists of 21 exhaustion items none of which refers to the work situation: e.g. 'being tired' and 'feeling weak' (physical exhaustion); 'feeling depressed' and 'feeling "burned out"' (emotional exhaustion); 'being unhappy' and 'feeling rejected' (mental exhaustion). The authors present the BM as an instrument for self-diagnosis and offer interpretations for test scores. However, the normative data of the BM (see Pines et al., 1981, pp. 202–206; Pines and Aronson, 1988, pp. 220–222) are of little value because of the lack of essential statistical information. Besides, these data stem from rather heterogeneous groups such as students, workshop participants, and respondents from different nations. For instance, a re-analysis of the data of the American samples (excluding students) reveals that workshop participants have significantly higher burnout scores than non-participants. Nevertheless, the test authors simply state, without elaboration, that if a score is higher than four then the individual had 'experienced burnout', whereas individuals with a score of between three and four should 'evaluate their priorities and consider possible changes' (Pines *et al.*, 1981, p. 28). If one wants nonetheless to use the BM-normative data for the purpose of comparison, we suggest consideration of only the American samples, excluding students and workshop participants.[2]

Internal consistency and stability

The BM appears to be a highly reliable instrument with internal consistency coefficients (α) exceeding .90. This is not very surprising as many items can be considered either synonyms (e.g. 'feeling "burned out"' and 'feeling rundown') or antonyms (e.g. 'being happy' and 'being unhappy'). The stability of the BM is relatively high, as indicated by test–retest coefficients (r) ranging from .89 to .66 across 1 and 4 month intervals, respectively (Pines and Aronson, 1988, p. 220).

Factorial validity

Empirical evidence challenges the one-dimensionality of the BM and suggests the existence of three different, reliable and interrelated, dimensions that do not concur with those that are included in the definition of burnout of the test-authors (Enzmann and Kleiber, 1989; Schaufeli and Van Dierendonck, 1993):

1   demoralisation (e.g. 'feeling depressed', 'feeling rejected');

2   exhaustion (e.g. 'feeling "burned out"', 'being tired');

3   loss of motive (e.g. 'being happy', 'feeling optimistic' – items are scored reversed).

However, the third dimension seems to reflect an artefact that results from inconsistent answering patterns of those persons who do not recognise the reversed scoring direction of the items (Enzmann *et al.*, 1998).

Construct validity

BM-scores are, as expected, positively related to various poor job characteristics such as lack of social support, lack of autonomy, lack of feedback, and lack of variety. Typically, however, correlations with these work features are moderate, but rarely exceeding values of .40 (Pines *et al.*, 1981, pp. 213–218). The construct validity of the BM is further strengthened by the fact that its scores are also related to behavioural indicators of work-related strain. For instance, in facilities for the developmentally disabled with higher mean burnout values, turnover rates were higher (49%) compared with facilities with lower mean burnout scores (17%) (Weinberg *et al.*, 1983). Not surprisingly, BM-scores are negatively related to self-reported satisfaction from work, from life, and from oneself (Pines *et al.*, 1981, p. 209) and positively related to poor self-reported physical health, and to on-duty physical symptoms such as headaches, loss of appetite, nervousness, backaches, and stomachaches (Pines and Aronson, 1988, p. 221). Typically, correlations with these measures of satisfaction and ill-health are moderate, ranging between .30 and .50. By contrast, the discriminant validity of the BM is rather poor with regard to measures of general psychological and physical strain. This led Schaufeli and Van Dierendonck (1993) to the following conclusion in their burnout validity study:

> Basically, this questionnaire indicates the individual's level of subjective distress. (p. 645)

As expected, correlations with other burnout measures, such as the MBI, are substantive indicating good convergent validity, particularly as far as emotional exhaustion and depersonalisation are concerned ($.50 < r < .70$) (Corcoran, 1986; Stout and Williams, 1983). Correlations with personal accomplishment, on the other hand, are much lower ($-.25 < r < -.30$).

Conclusion

The BM is a reliable and reasonably valid research instrument that indicates the individual's level of exhaustion which is not necessarily job-related. Its factorial validity is not beyond question, though: instead of one dimension that reflects various aspects of exhaustion, there appear to be two strongly related dimensions, exhaustion and demoralisation. A third dimension, entirely consisting of positively phrased items, is likely to be an artefact. The BM is unsuited for individual diagnostic purposes.

Maslach Burnout Inventory (MBI)

To date, the MBI is almost universally used as *the* instrument to assess burnout (see Chapter 4). The instrument was introduced in the early 80s (Maslach and Jackson, 1981a; 1981b), the second edition of the test manual was published 5 years later (Maslach and Jackson, 1986), and recently the third edition appeared (Maslach *et al.*, 1996).

As we have seen in the previous Chapter, the MBI test authors describe burnout as a three-dimensional syndrome that is characterised by emotional exhaustion, depersonalisation, and reduced personal accomplishment. Accordingly, three elements of our working definition of burnout are included in the MBI: mental exhaustion, dysfunctional attitudes and behaviours (i.e. depersonalisation), and reduced effectiveness. Instead of one composite burnout score, a score for each subscale is computed. In contrast to the BM, most MBI-items explicitly refer to the work situation (e.g. 'I feel used up at the end of the workday').

Initially, the test-authors restricted burnout to professionals who work with recipients in some capacity, that is, those who have direct personal contact with service recipients (e.g. patients, clients, students, inmates, pupils). However, the latest edition of the test manual includes, in addition to the traditional MBI-Human Services Survey (MBI-HSS) and the MBI-Educators Survey (MBI-ES), the MBI-General Survey (MBI-GS) (Schaufeli et al., 1996a). The MBI-GS can be used in any occupational context and includes three subscales (exhaustion, cynicism, and professional efficacy) that parallel those of the original MBI, except that items do not explicitly refer to working with people.

The three original MBI-dimensions were not deduced theoretically before the proper test construction commenced. Instead, an inductive approach was followed and the dimensions were labelled after analysing nearly 50 items. Initially, the MBI-HSS also included an involvement scale that was dropped in the second version of the test manual. Similarly, the original double rating of each item in terms of frequency and intensity was abandoned in favour of the former since both ratings were extremely highly correlated ($r > .80$). Items are scored on a seven-point rating scale with fixed anchors that range from 'never' to 'every day'. It takes approximately 7 minutes to complete the MBI.

Internal consistency

Internal consistencies of the three MBI-HSS/ES scales are satisfactory with values of α ranging from .70 to .90 (Maslach et al., 1996). Recently, Lee and Ashforth (1996) analysed 47 studies that included nearly 10 000 respondents and computed overall reliability coefficients for each subscale that were well within the range of the test manual: emotional exhaustion (.86), depersonalisation (.76), and personal accomplishment (.77). Occasionally, internal consistencies lower than .70 have been found with the depersonalisation scale (Schaufeli et al., 1993a). There are some indications that depersonalisation should be treated as a multidimensional construct that includes various aspects such as distancing, hostility, unconcern, and rejection (Garden, 1987). Thus, the occasional low internal consistencies for depersonalisation are not only due to the small number of items, as is usually assumed, but may also reflect conceptual problems. Internal consistencies of the MBI-GS are equally satisfactory, ranging from .73 (cynicism) to .91 (exhaustion) (Leiter and Schaufeli, 1996).

Test–retest reliability

MBI-scores seem to be rather stable over time. According to the test manual, test–retest coefficients of the MBI-HSS/ES range from .60 to .82 across short periods of up to 1 month and only drop slightly when longer periods of up to 1 year are considered ($.57 < r < .60$). Similar values were found for the MBI-GS. Test–retest coefficients range from .60 to .67 across a 1 year period. (Maslach *et al.*, 1996). Quite consistently, (emotional) exhaustion appears to be the most stable burnout dimension, whereas depersonalisation (cynicism) is the least stable dimension.

Factorial validity

The factorial validity of all three versions of the MBI has been confirmed by a number of recent studies that employed advanced statistical techniques: confirmatory factor analysis using linear structural equations modelling (MBI-HSS – Gold *et al.*, 1989; MBI-ES – Byrne, 1994; Schaufeli *et al.*, 1994a; MBI-GS – Leiter and Schaufeli, 1996). These studies show convincingly that the assumed three-dimensional structure indeed fits better to the data than, for instance, a one-dimensional structure. In other words, burnout as assessed with the MBI is a multi-dimensional construct.

As expected, the three dimensions are interrelated. Based on nearly 50 studies, Lee and Ashforth (1996) computed correlations between the MBI-HSS/ES subscales. Emotional exhaustion and depersonalisation are strongly related ($r = .52$), whereas personal accomplishment is moderately related to emotional exhaustion ($r = -.33$) and to depersonalisation ($r = -.36$). The interrelations of the MBI-GS show a slightly different pattern: cynicism is not only highly related to exhaustion ($.44 < r < .61$), but also strongly related to professional efficacy ($-.38 < r < -.57$) (Maslach *et al.*, 1996).

Construct validity

Results on the construct validity of the MBI-HSS/ES are numerous and on balance quite positive. The next chapter includes an abundant number of studies with the MBI that confirm expected relations with a wide array of variables including demographic characteristics, personality characteristics, work related attitudes, individual health outcomes, and organisational outcomes. In this section we narrow our focus and concentrate on the convergent and discriminant validity, respectively.

Convergent validity studies indicate that, to a large extent, the MBI-HSS/ES scales measure the same construct as do other burnout instruments such as the BM and the SBS-HP (see above). Typically, correlations of emotional exhaustion and depersonalisation with other burnout self-report indicators are high with *r*s exceeding .50, whereas correlations with personal accomplishment are somewhat lower (*r*s of about .30). Generally, MBI-scores are moderately related to information from external sources such as peer-ratings (*r*s of about .30). Once again, emotional exhaustion appears to be the best validated dimension. For instance, Lawson and O'Brien (1994) used behaviour observation to assess burnout among direct-care staff working with the mentally retarded and found a moderate correlation of emotional exhaustion only with a lack of positive interaction with clients. Obviously, there is some discrepancy between results obtained by self-reports and by other methods. It is likely that an artefact plays a role: at least some portion of the common variance between self-report measures of burnout must be attributed to the use of a common method. Thus, the convergent validity of the MBI is reasonably well demonstrated for emotional exhaustion and depersonalisation, but for personal accomplishment it seems to be insufficient.

Evidence for the predictive validity of the MBI was recently presented by Sixma *et al.* (1998). They found that among Dutch general practitioners, initial levels of emotional exhaustion predicted their dropping out of the profession in the next 5 years. Other independent predictors were: gender (female), work experience, and number of patients.

Investigations that assess the discriminant validity of the MBI-HSS/ES in relation to other concepts do not always yield encouraging results. First, all MBI-burnout dimensions correlate comparatively highly with measures of job satisfaction (see Chapter 4). Secondly, although all MBI dimensions (particularly emotional exhaustion) are related to self-reported physical and psychological distress symptoms, they can be validly discriminated from these symptoms (Schaufeli and Van Dierendonck, 1993). Correlations with objective health indicators are much lower (see Chapter 4). Finally, a review of 12 studies shows that emotional exhaustion is strongly related to depression, whereas correlations with both other dimensions are less substantial (see Chapter 4). Nevertheless, these studies suggest that burnout and depression are related but distinct concepts. Or, as Glass *et al.* (1993, p. 153) conclude:

> Burnout, it seems, cannot be dismissed as little more than a career-related form of depression.

Thus, results on the discriminant validity of the MBI-HSS/ES are quite consistent: emotional exhaustion is substantively related with job satisfaction, distress symptoms, and depression, whereas depersonalisation is somewhat less strongly related and reduced personal accomplishment is only weakly related to these constructs. Taken together, with the exception of emotional exhaustion, the discriminant validity of the MBI-HSS is reasonably well established.

Foreign language versions

Since the MBI is globally employed to measure burnout, some notes have to be made on the use of the instrument in non-English-speaking countries. Generally speaking, foreign language versions have similar internal consistencies and show similar factorial and construct validity as the original American version. For instance, the three factor structure of the MBI-HSS appeared to be invariant across Germany, the Netherlands and France (Enzmann *et al.*, 1995). Despite similar psychometric findings across countries, levels of burnout seem to differ systematically. For instance, European nurses (Schaufeli and Van Dierendonck, 1995), human services professionals (Söderfeldt *et al.*, 1996), and teachers (Van Horn *et al.*, 1997) show lower levels of emotional exhaustion and depersonalisation compared to similar North American samples. Later in this chapter, we will return to this issue that, perhaps, has to do with a particular response set. It could be speculated that compared with European subjects, Americans tend to respond more extremely to self-report questionnaires. Alternatively, it could also be speculated that different cultural values might play a role. For instance, North American society is more achievement-oriented and may therefore produce more stressful jobs that may increase emotional exhaustion. In addition, exhibiting particular burnout symptoms, notably depersonalisation, may probably be more socially accepted in the strongly individualised North American society than it is in Europe where sentiments of group solidarity play a more significant role. Finally, jobs are perhaps more stressful in the US than in Western Europe.

Quite remarkably, the only European country where similar burnout levels as in the United States have been observed is Poland, where working conditions are relatively poor compared with Western European standards (Schaufeli and Janczur, 1994). At any rate, it is unlikely that the translation of the inventory can be held responsible for the differences in levels of burnout between North America and Europe, since lower burnout scores were found among nurses in English-speaking countries (the United Kingdom and Ireland) and higher scores were found in French Canadian nurses (see Schaufeli and Janczur, 1994, for a review).

Conclusion

The conclusion concerning the psychometric quality of the MBI is mixed. On the positive side, the factorial validity and the convergent validity as well as the reliability of the instrument are quite encouraging. This applies likewise to several non-English versions. On the other hand, emotional exhaustion in particular overlaps with related concepts such as depression and job satisfaction, as well as with distress symptoms. Therefore, it seems that that the most robust and reliable subscale, emotional exhaustion, that also displays the strongest convergent validity is at the same time the least specific dimension of burnout.

3.3 DIAGNOSTIC ASSESSMENT OF SEVERE BURNOUT

'Burnout' is not a formal, officially-known diagnostic label. However, although the syndrome is described in most handbooks of Health Psychology and Industrial and Organisational Psychology, it still does not appear in similar books on Abnormal Psychology or Psychiatry. The reason is that the latter books are usually based on officially recognised and accepted diagnostic classification systems, most notably the *Diagnostic and Statistical Manual of Mental Disorders (4th ed.)* (DSM-IV), issued by the American Psychiatric Association (1994) and the *International Classification of Mental and Behavioural Disorders* (ICD-10), issued by the World

Health Organisation (1992). Burnout is included in neither diagnostic system. However, this might change quickly as was predicted in an editorial of *The Lancet*:

> 'Burnout', (. . .) in the zeal for identification of new syndromes in psychiatry, may well become a formal diagnosis in future versions of the DSM classification of psychiatric disorders. (*The Lancet*, 1994, p. 1583)

Pending an officially sanctioned psychiatric status, we will examine how the individual assessment of burnout can be performed by using or adapting existing diagnostic tools: DSM-IV, ICD-10, and the MBI.

Burnout as Adjustment Disorder (DSM-IV)

A group of Canadian psychiatrists has argued that, in fact, burnout can be considered a particular mental adjustment disorder as described by DSM[3] (Bibeau *et al.*, 1989). Hence, in their view the term burnout is superfluous. According to DSM-IV (p. 623) an adjustment disorder is characterised by:

> . . . the development of clinically significant emotional or behavioural symptoms in response to an identifiable psychosocial stressor or stressors. The symptoms must develop within 3 months of the onset of the stressor(s). The clinical significance of the reaction is indicated either by marked distress that is in excess of what would be expected given the nature of the stressor, or by impairment in social or occupational (academic) functioning . . . By definition, an Adjustment Disorder must resolve within 6 months of the termination of the stressor.

Six subtypes of Adjustment Disorders are distinguished, depending on the predominance of particular symptoms, e.g. depressed mood, anxiety, or disturbance of conduct. According to Bibeau *et al.* (1989), the unspecified subtype comes closest to burnout. DSM-IV (p. 624) states:

> This subtype should be used for maladaptive reactions (e.g. physical complaints, social withdrawal, or work or academic inhibition) to psychosocial stressors that are not classifiable as one of the specific subtypes of Adjustment Disorder.

In addition, the Canadian psychiatrists propose somewhat more specific diagnostic criteria for burnout that fit within the larger framework of an unspecified Adjustment Disorder:

- severe fatigue – the principal subjective indicator of burnout.

This should be accompanied by:

- loss of self-esteem resulting from a feeling of professional incompetence and profound job dissatisfaction;
- multiple physical symptoms of distress without an identifiable organic illness;
- problems in concentration, irritability, and negativism;
- a significant decrease in work performance over a period of several months – the principal objective indicator of burnout.

Bibeau and his colleagues also mention the following four exclusion criteria. First, subjective and objective indicators of burnout should *not* result from sheer incompetence, which means that work performance used to be adequate and has declined since. Second, the individual should *not* suffer from major psychopathology. Third, family-related problems should *not* be responsible for the occurrence of burnout. Finally, severe fatigue resulting from monotonous work or a high work load is excluded because this is not necessarily accompanied by feelings of incompetence or lowered productivity.

This set of diagnostic criteria agrees remarkably well with our previous working definition of burnout, except that no preconditions such as frustrated intentions are specified and that the self-perpetuating nature of burnout is not considered. However, despite these similarities, it seems problematic to consider burnout an Adjustment Disorder under DSM because such disorders occur by definition as a more or less immediate response to an identifiable stressor. Although a discrete event, such as a blocked promotion prospect (see Box 2.1), may provoke burnout, it is not its prime 'cause' as in other maladaptive reactions that may occur after stressful life events such as unemployment, divorce, or death of a spouse. In such cases there is an identifiable stressor with an onset and an offset that is clearly causally related to the adjustment disorder. Recently, Wheaton (1996, pp. 44–47) distinguished between such 'event stressors' (i.e. discrete, observable stressful life events) and 'chronic stressors' (i.e. slowly and insidiously developing problematic conditions) as the two opposite poles of the stress continuum. Clearly, burnout results from chronic stressors. Typically, it is characterised by a rather slowly unfolding process of energy depletion that may take years and that gradually develops in response to a rather fuzzy set of job-related stressors. Burnout, by its very nature, is not a short-term response to a clearly identifiable stressor that dissolves after 6 months, rather it is a long-term process of mental erosion without a clearly identifiable cause.

Burnout as work-related Neurasthenia (ICD-10)

The ICD-10 Neurasthenia diagnostic label, on the condition that it is work-related, can be used for the individual assessment of burnout. Unlike DSM Adjustment Disorder, ICD-10 Neurasthenia neither poses particular time constraints on the disappearance of the symptoms, nor does it require an identifiable cause. According to the ICD-10 guidelines a formal Neurasthenia diagnosis requires (pp. 192–193):

■ either persistent and distressing complaints of feelings of exhaustion after minor mental effort, or persistent and distressing complaints of feelings of fatigue and bodily weakness after minimal physical effort;

■ at least one of the following six distress symptoms: muscular aches and pain, dizziness, tension headaches, sleep disturbance, inability to relax, or irritability;

■ that the patient is unable to recover from the symptoms by means of rest, relaxation or entertainment;

■   that the duration of the disorder is at least three months.

■   that the criteria for any more specific disorders do not apply.

These criteria cover, to a large extent, the elements of burnout that were identified in our previous working definition. However, no preconditions are distinguished, and dysfunctional attitudes and behaviours at work, as well as a sense of reduced effectiveness are not included. This is due to the fact that the context in which Neurasthenia occurs is not specified. For that reason, it is deemed necessary to add the criterion of work-relatedness, so that *work-related* Neurasthenia may be considered the most appropriate formal psychiatric label for burnout. Compared with the DSM Adjustment Disorder this label is more appropriate since it recognises the chronic nature of burnout.

A recent study showed that out-patients from a psychotherapeutic treatment centre who were diagnosed as suffering from work-related Neurasthenia had significantly higher scores on MBI emotional exhaustion and depersonalisation compared with those who did not receive that diagnosis but suffered from various other mental disorders (e.g. mood disorder, post-traumatic stress disorder, social phobia) (Schaufeli *et al.*, in press). This result suggests that the formal ICD-10 Neurasthenia diagnosis, to which the criterion of work-relatedness is added, taps the notion of burnout as measured by two of the three scales of the MBI. Quite remarkably, scores on the BM did not differ significantly between both out-patient groups. It is most likely that this is caused by the fact the BM reflects exhaustion *per se*, rather than work-related exhaustion. Obviously, all out-patients are likewise exhausted, but only among the 'burned out' patients is exhaustion work-related.

Concerning the use of the MBI for assessment of clinical burnout

Principally, the MBI can be used to assess the individual's level of clinical burnout. However, the current classification of burnout levels is based on arbitrary statistical norms: the MBI-HSS normative sample has been divided into three equally sized groups of 33.3% each, assuming that the top-third experiences 'a high degree of burnout', the middle-third experiences 'an average level of burnout', and the bottom-third experiences 'a low level of burnout'. Accordingly, a cut-off point is determined that serves as a dividing line between those who are burned out and those who are not: that is, a score that corresponds with the top-third of the normative distribution of each of the MBI subscale scores. It is important to note that no clinically valid reason exists for using the top-third/bottom-third split as the dividing line between clinical cases and non-cases. It is an arbitrary statistical criterion, rather than a criterion that is based on clinical experience. The test-authors indicate that the categorisation of burnout into three levels is intended as feedback for individual respondents. They rightfully warn that:

> . . . neither the coding nor the original numerical scores should be used for diagnostic purposes, there is insufficient research on the pattern(s) of scores as indicators of individual dysfunction or the need for intervention. (Maslach *et al.*, 1996, p. 9)

Recently, for the Dutch, MBI-HSS clinically valid cut-off points were proposed (Schaufeli *et al.*, in press). Out-patients who received a work-related Neurasthenia diagnosis showed mean emotional exhaustion and depersonalisation scores that exactly corresponded with the 95th and 75th percentiles of the Dutch normative MBI-HSS sample, respectively. Accordingly, an empirically derived qualification could be attached to the percentile scores of the Dutch normative working sample: 'very high' levels of emotional exhaustion (i.e. scores exceeding the 95th percentile) and 'high' levels of depersonalisation (i.e. scores exceeding the 75th percentile) are comparable with mean levels of clinically burned-out patients who seek treatment and satisfy the diagnostic criteria for work-related Neurasthenia.

It must be stressed that such clinically validated cut-offs are nation-specific and should not be used outside the country they were developed in. In other words, it does not make sense to use cut-off points that are obtained in one country to classify individuals in another country. For instance, the use of the Dutch cut-off points would lead to flagrant misinterpretations in American samples since, generally, burnout levels are higher in the US than in the Netherlands. On the other hand, if American cut-off points had been used in the Dutch provisional normative MBI-HSS sample, a dramatic underestimation of emotional exhaustion and depersonalisation rates would have occurred (both 13% instead of about 33%), whereas the rate of those low on personal accomplishment would have been likewise overestimated (55% instead of about 33%) (Schaufeli and Van Dierendonck, 1995). In addition, the results of the Dutch study on the clinical validity of the MBI-HSS suggest that taking the top-third as a cut-off, as is proposed in the American test manual, is likely to severely overestimate the number of burned-out cases. Using this quite liberal criterion, too many individuals are considered 'burned out'. Accordingly, a more conservative cut-off, such as the 75th or 95th percentile, would be much more appropriate.

At least two studies show that the majority of individuals who score highly on the MBI were considered 'psychiatric cases' according to criteria of the General Health Questionnaire, an instrument that is used in epidemiological studies to detect minor psychiatric disorder. For instance, 52% of those who scored above the median on all three MBI dimensions simultaneously were identified as such cases (Golembiewski *et al.*, 1992). When a more conservative MBI cut-off point, the 95th percentile, is considered the picture becomes even more dramatic (Schaufeli and Van Dierendonck, 1994). Of those who scored in the top 5% on emotional exhaustion and depersonalisation, 70% were identified as psychiatric cases. For reduced personal accomplishment this is only 14%. These results confirm that a conservative cut-off is more appropriate in terms of classifying 'real' burnout cases.

Conclusion

To date, no formal diagnostic classification system includes 'burnout'. The existing ICD-10 diagnostic guidelines for Neurasthenia, if work-related, come closest to our working definition of burnout. Preliminary evidence suggests that the work-related Neurasthenia diagnosis is valid. Alternatively, the MBI can be used for the purpose of individual assessment of burnout, provided that its cut-offs are clinically valid-ated, which, to date, seems to be the case only in the Netherlands. Among those who score highly on the MBI, many need professional help. Therefore, it seems important that 'burnout' is recognised and officially sanctioned as a formal medical diagnosis so that employees can legally pursue their mental health claims and seek treatment. This is largely a political decision that must be based on sound scientific evidence of the validity of a particular set of diagnostic criteria. The criteria as specified by the Canadian group might be a good starting point for such an endeav-our. However, for the time being, ICD-10 work-related Neurasthenia seems to be the best formally recognised alternative. We are fully aware of the danger that a formal burnout diagnosis might foster further medicalisation and individualisation, which might increase the stigmatising burden. However, we feel that a formal recognition that opens the way for treatment and mental health claims for those who suffer from burnout outweighs this potential disadvantage.

3.4 PREVALENCE

How often does burnout occur? The answer to this rather simple question is not easy, given the present state of the art. Strictly speaking, we do not know. In order to answer this question properly, a measurement instrument is needed that discrimin-ates between burned-out cases and non-cases. But as we have seen in the previous section, such instruments are not available. Only the Dutch MBI-version has provi-sional, clinically validated cut-off points. Based on these cut-offs it is estimated that, depending on the MBI-dimension, between 3% and 16% of the Dutch human services professionals suffer from severe clinical burnout (Schaufeli and Van Dierendonck, 1995). However, these cut-offs are preliminary and further research is needed. The next best option would be to use statistical norms that allow one to classify respondents as being high or low on burnout, relative to some representat-ive sample. Although the normative data of the MBI are based on numerous studies and thousands of respondents in the United States, we do not know whether they are truly representative for a certain profession or occupational group because the data are not based on random samples of the respective groups. To complicate things even further, instead of a single burnout score, the MBI assesses three dimen-sions of burnout that can only be collapsed into one composite score at the expense of a substantial loss of information. Consequently, if we use the normative data of the MBI manual, the simple question 'How often does burnout occur?' cannot be answered.

A possible way out, at least as far as the dimensionality of burnout is concerned, has been proposed by Robert Golembiewski and his colleagues. They computed a single burnout score that is not based on the top-third criterion as suggested by the MBI test-authors, but on the median, being a less conservative cut-off point. Con-sequently, by definition, 50% of the individuals in their normative sample score highly on one of the three dimensions. According to Golembiewski and his col-leagues, the most severe phase of burnout is characterised by high scores on all three MBI-dimensions (of course, for this purpose personal accomplishment is reversed into reduced personal accomplishment – for details see Chapter 5). In their most recent book, more than 60 North American studies are summarised that include almost 25 000 subjects ranging from accountants and managers, through nurses, clerks, and production workers, to students of police academies and MBA programmes (Golembiewski *et al.*, 1996, pp. 159–163). Overall, 20% are categorised in the most severe phase of burnout, with percentages ranging from study to study between 0% and 35%. The majority of the studies report burnout rates between 15% and 25%. These results lead the authors to conclude:

> Whatever one's tolerable level of burnout in organisations, not to mention one's ideal, the estimates implied (. . .) are far too high for anyone's comfort.   (p. 165)

In addition, they list more than 20 studies including approximately 7000 subjects that were carried out outside North America in countries like Japan, China, Belarus, Poland, Ghana, and Saudi Arabia (pp. 169–170). Burnout rates are even higher in most of these samples. Overall, 37% of the subjects were categorised in the most severe phase of burnout, ranging in various samples between 12% and 69%. These figures must be interpreted with caution since they are based on US cut-offs. When local cut-offs are employed the picture changes dramatically. For instance, according to local norms, 24% of the Japanese employees were suffering from severe burnout against 68% when US cut-offs were employed! (p. 181). Moreover, the US normative sample, which consists of 1535 employees from a US federal agency, is unlikely to be representative of the nation's workforce. Thus, information on the prevalence of burnout provided by Golembiewski and his colleagues is not likely to be accurate because their normative sample is not representative. Besides, they use an arbitrary, non-validated and rather liberal inclusion criterion for burnout cases.

Since, obviously, the question 'How often does burnout occur?' cannot be answered, we seek to answer two more specific questions about the relative prevalence of burnout: (1) Does the level of burnout, as measured with the MBI, differ between various occupational fields and professions?; (2) Can such differences be observed across nations? In order to answer these questions we analysed two data-sets, respectively: (1) A batch of 73 US studies including over 17 000 respondents that were published between 1979 and 1998; (2) Compiled data from 57 American samples $(n = 12\ 239)^4$ and 27 Dutch studies $(n = 10\ 502)$ – two countries with the largest MBI data-bases.

Are there specific burnout profiles in various occupational fields and professions?

Table 3.1 summarises the mean MBI burnout levels based on 73 US studies according to six occupational fields that are included in the test manual (i.e. teaching, post-secondary education, social services, medicine, mental health, and others) (Maslach *et al.*, 1996). In addition, burnout levels of various professions are displayed.

Table 3.1 Normative data of the MBI based on 73 US studies published between 1979 and 1998

Field/Profession	Emotional exhaustion				Depersonalisation				Reduced personal accomplishment			
	samples	n	mean	sd	samples	n	mean	sd	samples	n	mean	sd
teaching	**6**	**5481**	**28.15**	**11.99**	**6**	**5481**	**8.68**	**6.46**	**7**	**6191**	**11.65**	**7.41**
higher education	**5**	**877**	**19.17**	**10.72**	**5**	**877**	**6.02**	**5.56**	**5**	**877**	**10.49**	**6.77**
social services	**7**	**1631**	**24.29**	**12.79**	**7**	**1619**	**9.47**	**7.16**	**7**	**1601**	**13.45**	**8.55**
social worker	6	628	20.82	10.17	6	628	6.94	5.60	6	628	12.85	7.80
unspecified	1	1003	26.47	13.75	1	991	11.08	7.56	1	973	13.84	8.98
medicine	**14**	**2021**	**23.86**	**11.57**	**14**	**2021**	**7.95**	**6.47**	**14**	**2016**	**12.38**	**7.96**
nurse	11	1542	23.80	11.80	11	1542	7.13	6.25	11	1541	13.53	8.15
physician	3	479	24.03	10.77	3	479	10.59	6.46	3	475	8.64	5.93
mental health	**19**	**2290**	**20.42**	**10.10**	**19**	**2290**	**6.29**	**4.72**	**19**	**2290**	**8.95**	**6.99**
psychologist	6	1382	19.75	9.77	6	1382	6.14	4.45	6	1382	7.06	6.07
counsellor	4	422	20.52	8.97	4	422	6.64	4.26	4	422	12.65	6.75
staff	6	333	22.09	10.96	6	333	6.03	5.22	6	333	11.24	7.58
unspecified	3	153	22.61	12.96	3	153	7.31	6.64	3	153	10.84	7.99

Table 3.1 (Cont'd)

Field/Profession	Emotional exhaustion				Depersonalisation				Reduced personal accomplishment			
	samples	n	mean	sd	samples	n	mean	sd	samples	n	mean	sd
other	**21**	**5541**	**20.64**	**11.00**	**20**	**5010**	**8.02**	**6.36**	**18**	**4270**	**12.64**	**7.98**
police officer	2	430	17.55	10.90	2	430	12.48	7.22	2	430	15.94	10.19
probation/correction officer	2	386	19.49	11.33	2	330	9.73	7.01	2	322	15.14	8.37
librarian	2	609	20.22	10.75	2	609	8.30	5.96	2	609	11.39	6.83
senior executive	1	224	15.10	10.50	1	224	5.80	4.20	1	224	12.30	7.50
employee	5	1780	22.29	11.33	4	1317	8.78	6.85	4	1317	11.94	7.56
missionary/pastor	3	852	23.39	10.01	3	840	8.74	5.14	1	108	9.70	6.50
student	3	229	19.81	10.61	3	229	5.15	5.64	3	229	14.68	7.63
unspecified health	3	1031	18.85	10.28	3	1031	4.99	4.85	3	1031	12.03	7.66
total	**72**	**17 841**	**23.54**	**11.91**	**71**	**17 298**	**8.03**	**6.33**	**70**	**17 245**	**11.73**	**7.75**

Notes: The weighted averages of coefficients α are .86, .72, and .76 for emotional exhaustion, depersonalisation, and personal accomplishment, respectively (17 studies, $n = 4567$). The formula to recode the means of reduced personal accomplishment (rPA) to personal accomplishment (PA) is: PA = 48 − rPA.

It appears that, compared with the MBI test manual, levels of emotional exhaustion of the studies summarised in Table 3.1 are significantly higher for all occupational fields (total score: 23.54 versus 20.99). This was particularly true for those studies that have been published since 1987. Perhaps, overall job demands have increased during the past decade. In contrast, levels of depersonalisation and reduced personal accomplishment[5] are on balance rather similar to the test manual, although significant differences exist between specific occupational fields.

As far as emotional exhaustion is concerned, levels are clearly highest among teachers. Intermediate levels are found in the social services and in medicine, whereas workers in mental health and post-secondary education experience the lowest levels of emotional exhaustion. For depersonalisation the picture is slightly different. Social workers and teachers report the highest levels, whereas levels in post-secondary education and in mental health are lowest. Two professions exhibit particularly high levels of depersonalisation: physicians and police officers. Perhaps this reflects their occupational socialisation that is characterised by objectiveness and distance (e.g. the prototypical 'John Wayne syndrome' in police officers). An alternative explanation would be that these are typically male-dominated professions with relatively few women in them. It is known that, as a rule, men experience more depersonalisation than women (see Chapter 4). Finally, reduced personal accomplishment is especially found in the social services and among nurses, police officers, and probation/correction officers. Not surprisingly, the most highly-trained professionals (i.e. physicians and psychologists) experience the strongest sense of accomplishment in their jobs. Particularly in the field of medicine, large differences exist between professions. Physicians and nurses experience about the same level of emotional exhaustion, but the former have much higher scores on depersonalisation, whereas the latter are characterised by a strongly reduced sense of personal accomplishment.

Do burnout profiles differ across nations?

In the previous section we have seen that burnout profiles systematically differ across occupational fields. For instance: teachers are characterised by relatively high levels of emotional exhaustion; in the social services, levels of depersonalisation and reduced personal accomplishment are particularly high; and law-enforcement is characterised by low levels of emotional exhaustion and high levels on both other burnout dimensions. Since we only examined American studies these profiles could be nation-specific. If similar profiles could be observed in another country, they would be likely to be occupation specific. Thus, we compared burnout profiles of the US occupational fields with similar fields in the Netherlands.

First of all, in the United States, emotional exhaustion and depersonalisation are significantly higher than in the Netherlands, whereas reduced personal accomplishment is significantly lower. As we have seen before, the former is in agreement with several other European studies. It is important to note that these differences are found in virtually every occupational field, except in mental health where no differences between the two countries were observed. These results once more underscore the necessity to employ nation-specific norms.

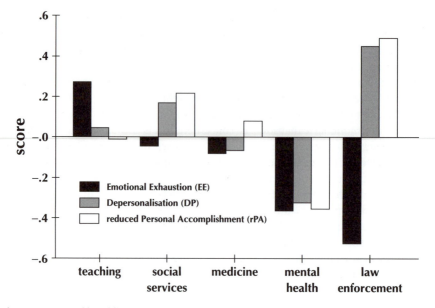

Figure 3.1 Profile of burnout by occupational fields (z-scores, US)
Notes: Accumulated from 57 studies; teaching: $n = 5481$; social services: $n = 1631$;
medicine: $n = 2021$; mental health: $n = 2290$; law enforcement: $n = 816$.

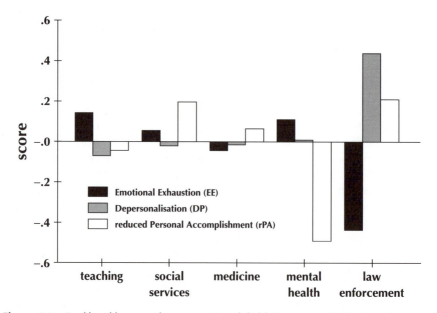

Figure 3.2 Profile of burnout by occupational fields (z-scores, Netherlands)
Notes: Accumulated from 27 studies; teaching: $n = 2284$; social services: $n = 459$;
medicine: $n = 6287$; mental health: $n = 1019$; law enforcement: $n = 453$.

Figures 3.1 and 3.2 show profiles of the MBI-dimensions for five occupational fields in the United States and in the Netherlands, respectively. Scores are standardised to a mean of 0.0 and a standard deviation of 1.0 within each nation, thus, differences of burnout levels between countries cannot be observed. The profiles in both figures must be interpreted relative to the nation's average: positive values indicate that scores on that particular dimension are higher than the nation's average across all occupational fields, negative values indicate that scores are lower. Four conclusions can be drawn from a comparison of Figures 3.1 and 3.2. First, profiles of law enforcement are highly similar in both countries. They are characterised by comparatively high levels of depersonalisation and reduced personal accomplishment, and low levels of emotional exhaustion. Secondly, teaching and medicine partly show similar profiles. Teaching is characterised by the highest level of emotional exhaustion, with both other dimensions close to the nation's average. Medicine, in both countries, is characterised by somewhat low levels of emotional exhaustion and depersonalisation, and by slightly high levels of reduced accomplishment. Thirdly, the profiles of workers in social services and mental health show a clearly different pattern. In the United States, levels of depersonalisation in the social services are relatively high, whereas they are about average in the Netherlands. The largest differences are observed among mental health workers though. In the United States they experience quite low levels of emotional exhaustion and depersonalisation, whereas in the Netherlands levels are relatively high (emotional exhaustion) or average (depersonalisation).

In conclusion, it seems that despite absolute differences in levels of burnout between both countries, similarities in burnout-profiles are larger than differences. The fact that the most striking similarities are found in law enforcement and the largest differences in mental health may be caused by the heterogeneous composition of professions in the former group compared with the latter. For example, burnout profiles of psychiatric nurses and psychiatrists are more likely to be different than profiles of police officers or correctional officers.

Conclusion

To date, it is not possible to accurately estimate the prevalence of burnout. However, using the MBI, a relative comparison can be made between levels of burnout across various occupational fields and professions. This reveals specific patterns for each burnout dimension: relatively high levels of emotional exhaustion are found in teaching and social work, and among physicians and nurses; high levels of depersonalisation are found in social work and law enforcement, and among physicians; high levels of reduced personal accomplishment are observed in the social services and in law enforcement and among nurses. Moreover, it seems that burnout profiles for particular occupational groups (teaching, law enforcement, and, to a somewhat lesser degree, medicine) are consistent across nations. The fact that such profiles do exist underscores the validity of the multidimensional perspective on burnout. It strongly argues against the use of a single burnout score, which would obscure important differences between groups.

3.5 SUMMARY

The methods that have been used so far to assess burnout are rather one-sided – almost exclusively easy to administer self-report questionnaires. The psychometric quality of two of these questionnaires has been more or less extensively studied. It appears that conceptually, as well as technically, the MBI is superior to the BM. By contrast with the BM, the MBI is multidimensional and its items are job-related, therefore this questionnaire is more specific. Besides, most of its psychometric features are quite encouraging (i.e. reliability, test–retest correlation, factorial validity, and convergent validity). Only its discriminant validity is relatively poor. This is particularly true for emotional exhaustion that is quite strongly related to, for instance, job satisfaction, psychosomatic symptoms, and depression.

Although no formal burnout diagnosis exists, it seems that clinical forms of burnout are best covered by the existing ICD-10 diagnostic guidelines for Neurasthenia, if work-related. Potentially, the MBI can also be used to assess such severe forms of burnout (i.e. work-related Neurasthenia), provided that clinically validated cut-offs are available. To date, this seems only to be the case for the Dutch version of the MBI.

Since neither a clinically valid nor a statistically valid criterion for burnout exists, its prevalence cannot be adequately assessed in absolute terms. Thus, we don't know how many individuals are actually 'burned-out'. Instead, using large US and Dutch MBI data-bases, we compared the relative occurrence of burnout symptoms in various occupational fields and professions. Particular characteristic profiles could be identified that seemed to be rather consistent across nations. For instance, teachers had the highest emotional exhaustion levels, whereas physicians and law enforcement personnel reported the highest levels of depersonalisation, and reduced personal accomplishment was most common in the social services and in law enforcement.

Notes

1   Other multi-dimensional questionnaires that have been occasionally used: the Burnout Assessment Inventory (BAI) (Clouse, 1982); the Emener-Luck Burnout Scale (Emener *et al.*, 1982); the Teacher Attitude Scale (Farber, 1984); the Teacher Stress Inventory (Fimian, 1984); the Nursing Stress Scale (Gray-Toft and Anderson, 1981); the Psychologist's Burnout Inventory (Ackerley *et al.*, 1988); and the Medical Personnel Stress Survey (Hammer *et al.*, 1985). Additional one-dimensional questionnaires that basically assess exhaustion have also been used: the Energy Depletion Index (Garden, 1987); the Meier Burnout Assessment (Meier, 1984).

2   Based on 10 studies ($N = 1535$), assuming a normal distribution of the scores and an internal consistency of $\alpha = .92$, US-norms are: mean = 3.12, standard deviation = 0.77, standard error of measurement = 0.22, lower third < 2.78, top third > 3.45 (computed by the authors; cf. Pines *et al.*, 1981, pp. 202–206).

3   Bibeau *et al.* (1989) referred to the previous DSM-III version. Meanwhile, the next updated version, DSM-IV, was published in 1994.

4   All 57 samples were taken from the first batch, but because a similar categorisation had to be used as that used in the Dutch data, some studies were dropped. For instance, in the Netherlands no MBI data are available for post-secondary educators so that the American studies that include this occupational field were excluded.

5   To facilitate the interpretation of personal accomplishment as a dimension of burnout we have recoded the scores of 'personal accomplishment' as 'reduced personal accomplishment' throughout the chapter.

CHAPTER FOUR

What are the correlates, causes and consequences?

Empirical research

Over the years, research interest in burnout has grown immensely. But what does this impressive body of research tell us about the possible correlates, causes and consequences of burnout? This chapter reviews the empirical evidence on this issue. First, however, we will discuss some quantitative and qualitative trends in burnout research over the past two decades. This overview of the quantity and quality of burnout research serves as a necessary background for reviewing the empirical work in subsequent sections.

4.1 CONCERNING QUANTITY AND QUALITY

How many studies on burnout have been conducted in the past two decades? What burnout instruments have been used by researchers? What occupational groups have been studied most frequently? These three questions are answered in the following sections. Following that, some critical notes on the quality of current burnout research will be made.

How many burnout studies are there?

A huge number of studies exist containing the key word 'burnout' in the title. A search in databases[1] and in specialised bibliographies[2] renders more than 5500 entries. The bulk of this literature (61%) has been published in professional journals, about 17% are dissertations, 10% are book chapters or books, and the remaining 12% are research documents, conference papers or master's theses.

Between 1975 and 1980, the yearly number of publications on burnout increased steadily from 5 to over 200 (see Figure 4.1). Since 1980 the average publication rate has been about 200 per year, with an additional increase to 300 at the end of the

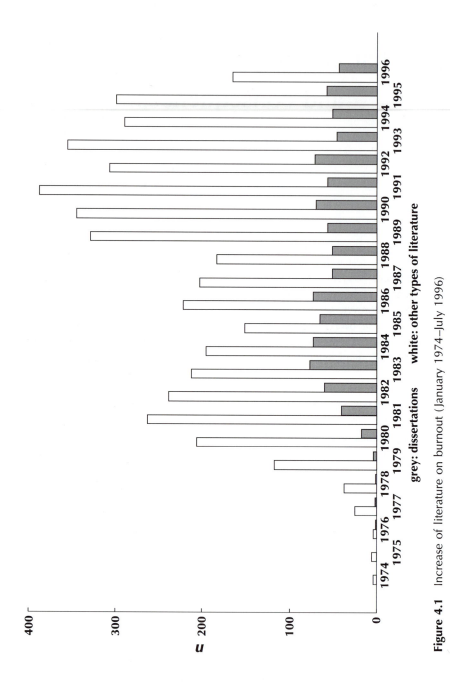

Figure 4.1 Increase of literature on burnout (January 1974–July 1996)

1980s. These impressive figures should not be misinterpreted, though, because this remarkable increase is not unique to burnout: it is also observed for publications on occupational stress. Hence, it looks like similar scientific and societal developments stimulated the interest in burnout as well as in occupational stress (see Chapter 1).

It is difficult to estimate how much of the literature actually represents empirical research. Until at least 1981, most journal articles were of an anecdotal and descriptive nature. However, with the availability of easy-to-administer self-report instruments such as the BM and the MBI (see Chapter 3), quantitative research increased considerably. This is illustrated by the growing number of articles listed in the *Psychological Abstracts* in which the term 'burnout' is mentioned, as well as by the fact that since 1981 burnout has been a topic of more than 900 doctoral dissertations which are almost exclusively empirical in nature. Figure 4.1 shows that since the mid 1980s there has been a steady output of over 50, mostly American, dissertations per year.

Which instruments are used?

The dominance of the MBI as the instrument of choice to assess burnout is clearly illustrated in Table 4.1. The data displayed are restricted to doctoral dissertations because, in contrast to journal articles, their abstracts mostly refer explicitly to the measurement devices used. Table 4.1 shows that, in the dissertations that are indexed in the Dissertation Abstracts International and that were written between 1976 and 1996, the MBI was employed ten times more often than all other burnout measures together! Or, stated differently, of all burnout measures the MBI was employed in 91% of the cases.

A remarkably similar picture emerges if we look at other publications, such as journal articles. Based on 498 publications in which one of the three most prominent burnout instruments were used (not including publications of the test authors themselves) 93% refer to the MBI, 5% to the BM, and 2% to the SBS-HP. These

Table 4.1   Burnout measures used in dissertations 1976–1996 (*n* = 963)

Measure	frequency of use	%
MBI	581	59.7
BM	25	2.6
SBS	19	2.0
BAI	11	1.1
FBS	1	0.1
not mentioned	337	34.6

Notes: **MBI** = Maslach Burnout Inventory; **BM** = Burnout Measure;
SBS = Staff Burnout Scale; **BAI** = Burnout Assessment Inventory;
FBS = Freudenberger Burnout Scale.

results clearly underscore that the MBI is the instrument of choice to measure burnout. No wonder that the definition of burnout has become equivalent to the way it is measured by the MBI!

Therefore, almost inevitably, we will restrict ourselves in this Chapter to studies that employed the MBI. It is important to note that the operationalisation of burnout as implied in the MBI is quite narrow compared with the broad range of symptoms and definitions that was presented in the preceding chapters. On the other hand, restricting ourselves to studies that used the MBI has the advantage that results are comparable since they are based on a common notion of burnout.

Which occupations are studied?

Figure 4.2 presents the occupational fields in which burnout, as measured with the MBI, has been studied. It shows that burnout is predominantly researched in the health and teaching professions. Other occupational fields are social work, administration and management, and law enforcement – police, correction and probation officers. The largest homogeneous occupational group is teachers, who represent 22% of all samples studied or 79% of the 'teaching and education' category. The largest occupational groups in health care are nurses (17% of all samples) and psychologists or psychotherapists (4% of all samples). Other rather large occupational groups are social workers (7% of all samples), school principals (3%), special educators (3%), physicians (2%), correction and probation officers (2%), and

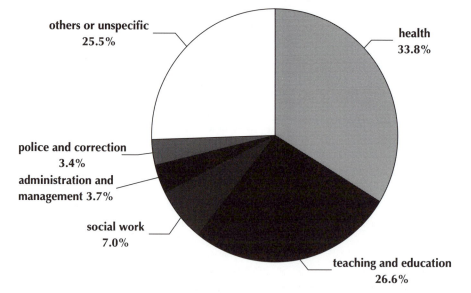

Figure 4.2 Occupational fields of samples studied with use of the MBI
Note: Based on 473 studies published in journal articles or books and 538 dissertations between 1978 and 1996

police officers (2%). The largest single group in the category 'others or unspecified' are pastors or missionary workers (2%). Thus, burnout research covers a wide range of human services professions.

Clearly, it is not just the significance of burnout in a particular field or profession that is responsible for the distribution that is displayed in Figure 4.2. It is likely that the availability of samples is also an important factor that plays a decisive role in the researcher's choice for a particular occupation. For instance, it is remarkable that dissertations very often include teachers or educators, whereas other studies predominantly include health care professionals. Furthermore, samples investigated in dissertations are often collected from university faculty or staff, not so much because university members are considered to be a high risk group for burnout but because sampling at universities is more convenient. Hence, the dominance of teachers, for instance, may somewhat overestimate the significance of burnout in this occupational field compared with health care and social work. It is important to note that the frequency distribution that is displayed in Figure 4.2 shows on which occupational fields our review of correlates, causes, and consequences of burnout is based.

Methodological quality

Despite the impressive quantity of empirical publications the quality is often questionable. Kilpatrick (1989) carefully evaluated the methodological rigour of nearly 200 empirical studies. Her observations confirm the one-sidedness and poor methodological quality of early burnout research: about 90% are surveys; 96% suffer from selection bias; 88% use non-random samples; 89% are one-shot designs; and response rates are rather moderate (mean rate 66%). On the other hand, she also noticed a promising trend:

> . . . burnout studies are increasing in methodological rigour, and a growing number of studies address interventions and are longitudinal.   (p. 42)

Despite this optimistic note, almost 10 years later the overwhelming majority of burnout research still is descriptive and exploratory in nature. From a methodological perspective there are at least five severe problems with such an approach. First of all, the research is usually based on *ad hoc* theories that only take into account direct effects (e.g. of job demands) despite the fact that the theoretical models of burnout are much more complex, assuming moderating and mediating factors with direct, indirect, reciprocal, immediate, or delayed effects (see Chapter 5).

Second, because data are usually exclusively based on self-reports it is easy to fall into the 'triviality trap', especially when self-report measures are used with overlapping item content. For instance, as indicators of workload, items have been used such as 'How great is the amount of emotional strain that nursing puts on you?' (Leiter, 1988) or 'My job requires working very hard' (Karasek, 1985), while the MBI includes items like 'Working with people all day is really a strain for me' or 'I feel I'm working too hard on my job'. No wonder that workload as operationalised in such ways correlates substantively with emotional exhaustion: burnout and its assumed antecedent are simply identical! In order to improve the validity of results and to avoid this triviality trap, objective measures should be included such as performance appraisals, registered absenteeism, or physiological parameters.

Third, in burnout research adequate sampling seems to be the exception rather than the rule. Generally, convenient samples rather than random and representative samples are investigated. However, adequate sampling is necessary in order to generalise research findings beyond the particular sample under study. A particular pitfall in burnout research is the 'healthy worker effect': the underestimation of the effects of burnout because exclusively working, and thus more or less healthy, employees are investigated. Because of this selection-effect our present knowledge of burnout is exclusively based on those who are in the *process* of burning out, rather than on individuals who have reached the end-state of that process and suffer from clinical burnout (i.e. work-related neurasthenia – see Chapter 3).

Fourth, response rates are often less than 50%. This impairs the internal validity of the results because self-selection of individuals may systematically obscure the importance of particular factors. For example, individuals who are under extreme time pressure might be more likely to refuse to participate. In that case, the role of time pressure in the development of burnout may be underestimated.

Fifth, often correlations between variables measured at one point in time are used for causal inferences about the antecedents and/or consequences of burnout. Accordingly, the basic fact that a causal effect can exclusively be observed after a particular time interval in which the assumed cause may exert its influence is ignored. In addition, the fundamental methodological premise that correlation co-efficients can only refute a causal relationship but never confirm it is neglected.

The final issue of causality is the most serious methodological objection that can be raised against much burnout research. Because of the present lack of empirically supported theories, one cannot rely on deductive reasoning that is grounded in an established theoretical framework. Therefore, it is all the more important that a sound methodological procedure that allows for valid causal conclusions is followed. Basically, only experiments and longitudinal studies allow hypotheses about causality to be tested and should thus be preferred to cross-sectional studies.

In the next sections, we will review the outcomes of over 250 studies on burnout that can be grouped according to antecedents or possible causes (Section 4.2) and concomitants or possible consequences (Section 4.3). The majority of these studies are based on cross-sectional data so that causal relationships cannot be confirmed. Accordingly, most studies report on assumed or possible causes and consequences, which are, as a matter of fact, correlates of burnout. In our review, we supplement and contrast cross-sectional findings with outcomes from longitudinal studies. Special attention is paid in Section 4.4 to longitudinal studies that investigate the relationship between job demands and burnout. The reason for this is twofold: (1) the practical and theoretical importance of the job demands-burnout relationship; (2) the availability of a reasonable number of methodologically sound longitudinal studies.

Table 4.2   Correlates of burnout: possible causes

biographical characteristics

age	– –
gender	+
work experience	–
marital status	–
level of education	(+)

personality characteristics

hardiness	– – –
external control orientation	+ +
confronting coping style	– –
self-esteem	– –
'feeling type'	+
Type A behaviour	+
neuroticism (anxiety)	+ + +
extroversion	–

work related attitudes

high (unrealistic) expectations	(+)

work and organisational characteristics

workload	+ + +
time pressure	+ + +
role conflict and ambiguity	+ +
hours worked	+
direct client contact	+ +
number of clients	+
severity of clients' problems	+
social support from colleagues or superiors	– –
lack of feedback	+ +
participation in decision making	– –
autonomy	–

Note: The number of minus or plus signs denotes the strength and the direction of the relationship with burnout.

4.2 ANTECEDENTS OR POSSIBLE CAUSES OF BURNOUT

Possible causes of burnout can be classified into personality characteristics, work-related attitudes, and work and organisational characteristics (see Table 4.2). In addition, biographical characteristics are included as possible causes, although it would be more accurate to label these as co-factors. Obviously, gender is not a 'cause' of burnout, but it may be linked to other factors such as role expectations, role taking, or 'feeling type' that may act as causal agents. Likewise, age is not a 'cause' of burnout but it may be related to age-dependent factors such as occupational socialisation.

The number of minus or plus signs in Table 4.2 denotes the strength and the direction of the relationship with burnout, based on three subjective criteria: (1) the number of studies that found clear evidence for the relationship; (2) the methodological quality of these studies; (3) the consistency of results across studies.

Biographical characteristics

Of all biographical characteristics, age is the most consistently related to burnout (Mor and Laliberte, 1984; Birch *et al.*, 1986; Poulin and Walter, 1993a). Among younger employees, burnout is observed more often than among those aged over 30 or 40 years. This is in line with the observation that burnout is negatively related to work experience. Burnout seems to occur rather at the beginning of the career. For instance, Maslach (1982b) found that psychiatric nurses burned out about $1^1/_2$ years after starting their careers, attorneys after 2 years, and social workers after 2 to 4 years. These are relatively early stages of a career. Similar periods were reported by Pines and Aronson (1988). Some authors interpret the greater incidence of burnout among the younger and less experienced in terms of reality shock (Künzel and Schulte, 1986) or as an indicator of an identity crisis due to unsuccessful occupational socialisation (Cherniss 1980a). The MBI manual (Maslach *et al.*, 1996) shows the decline of burnout symptoms with growing age or work experience for all three dimensions, but most clearly for depersonalisation and emotional exhaustion. However, a cautionary note should be made because a particular selection effect, survival bias, cannot be ruled out: those who burn out early in their careers are likely to quit their jobs, leaving behind the survivors who exhibit low levels of burnout.

The relationship of burnout with gender is not so clear-cut. Although there are some studies showing that burnout occurs more often among women than among men (Maslach and Jackson, 1981b; Büssing and Perrar, 1991; Poulin and Walter, 1993a), the opposite is also found (Price and Spence, 1994; Van Horn *et al.*, 1997). On balance, women tend to score slightly higher on emotional exhaustion, whereas men score significantly higher on depersonalisation. This can partly be explained by sex role-dependent stereotypes. For instance, it has been argued that men hold instrumental attitudes, whereas women are more emotionally responsive and seem to disclose emotions and health problems more easily (Ogus *et al.*, 1990). Another explanation may be that, due to additional responsibilities at home, working women experience higher overall workloads compared with working men. Workload, in its turn, is positively related to burnout, particularly to emotional exhaustion (see below).

As far as marital status is concerned, those who are unmarried (especially men) seem to be more prone to burnout compared with those who are married (McDermott, 1984; Maslach and Jackson, 1985; Raquepaw and Miller, 1989). Single people seem to experience even higher burnout levels than those who are divorced (Maslach *et al.*, 1996, p. 47). Possibly, other factors such as age or psychosocial problems that codetermine marital status may explain this difference which has also been found in studies on general health and well-being (Stroebe and Stroebe, 1991).

Some indications were found that those with a higher level of education were more prone to burnout than less educated employees (Mor and Laliberte, 1984; Birch *et al.*, 1986; Cash, 1988). Probably, the higher expectations of the more educated individuals are responsible for this observation. Another explanation might be that the more highly educated individuals more often gain positions with more responsibility (over other persons). However, data in the MBI manual (Maslach *et al.*, 1996) show a consistent trend only for depersonalisation, whereas emotional exhaustion is highest in the least educated (no or some college education) and in the most educated groups (postgraduate work). Reduced personal accomplishment shows yet another trend: it is highest in the middle group that finished college. These different patterns may be related to the different kinds of jobs these three educational groups hold (see Chapter 3).

The number of studies on remaining biographical characteristics such as ethnicity (Colegrove, 1983; Dames, 1983), parents' socioeconomic status (Dames, 1983), or income (Colegrove, 1983; Mor and Laliberte, 1984) is too small to observe clear trends.

Personality characteristics

The role of personality characteristics or traits in the development of burnout is a complicated issue. Suppose that a particular personality characteristic correlates with burnout, does that mean that this trait directly causes burnout? The answer is no, a high correlation does not necessarily imply causality. Individuals may put themselves in situations that match with their personality and such situations may foster burnout. For example, a person of the 'feeling type' may be inclined to choose a career as a nurse, which is known to be a stressful profession (Garden, 1989). Furthermore, personality characteristics can moderate the effect of stressful situations on burnout such that certain traits may buffer or, conversely, enhance negative outcomes. In other words, person and situation interact in complex ways, and as a result it is difficult to interpret the meaning of correlations of burnout with personality traits.

All in all, more than 100 studies on burnout include personality characteristics. However, we will discuss only the eight most frequently studied characteristics (see Table 4.2). Because of the importance of the so-called 'Big-Five' personality traits in modern psychology, a special paragraph is devoted to their relationships with burnout.[3]

A 'hardy personality' is characterised by involvement in daily activities, a sense of control over events, and openness to change. This personality trait is consistently related to all three dimensions of the MBI: the more hardy persons are less emotion-ally exhausted, less depersonalised and have stronger feelings of personal accom-plishment. Hardiness and burnout share about 10% to 25% of their variance, the highest correlations are found with emotional exhaustion (e.g. Nowack, 1986; Pierce and Molloy, 1990).

A person's 'locus of control' may either be external or internal: those with an external locus of control attribute events and achievements to powerful others or to chance, whereas those with an internal locus of control ascribe events and achieve-ments to their own ability, effort, or willingness to risk. Although somewhat less consistent, compared with internals, externals are more emotionally exhausted, deper-sonalised, and experience reduced feelings of personal accomplishment. A recent review of eleven studies showed that external locus of control explains about 10% of the variance of emotional exhaustion and about 5% of depersonalisation and reduced personal accomplishment (Glass and McKnight, 1996).

Similar results have been reported on 'coping styles' and burnout. A review of twelve studies revealed that those who are burned-out cope with stressful events in a rather passive, defensive way, whereas active and confronting coping is associated with less burnout (Enzmann, 1996). Both confronting coping and avoiding coping share about 5–10% of the variance of emotional exhaustion and depersonalisation. With respect to personal accomplishment, confronting coping explains about 15% of variance, whereas the relationship with avoidant coping is clearly weaker (less than 5%). The relatively strong relationship of confronting coping with personal accomplishment may be due to a process of mutual reinforcement: when a problem-focused strategy is used, this is likely to trigger feelings of accomplishment, and vice versa (Lee and Ashforth, 1996).

Two qualifying comments have to be made about the relationship between coping and burnout. First, some have argued that coping is a state, i.e. a situation specific response, rather than a personality trait that is stable across situations (Lazarus and Folkman, 1984). Following this logic, correlations between ways of coping and burnout would indicate that burnout is related to particular situations that are per-ceived as being stressful, which, in their turn, stimulate a person to cope in a particular way. For instance, a nurse who copes with the emotional demands at her job by avoiding personal contact with patients does not do so because she has an avoidant coping style but because it is an instrumental response in that particular situation: in a stressful close relationship with her boyfriend she might use a con-fronting coping strategy. Secondly, withdrawal from work or from clients which closely resembles avoidant coping has been regarded as a core symptom of burnout (Cherniss, 1980b). In a similar vein, depersonalisation is characterised by behavi-ours that include mental or behavioural withdrawal (Maslach, 1982b). Accordingly, avoidant coping and burnout partly overlap conceptually.

It has been argued that low levels of hardiness, an external locus of control, and an avoidant coping style typically constitute the profile of a stress-prone individual (Taylor and Cooper, 1989; Semmer, 1996). Obviously, results from burnout research confirm this personality profile.

In their meta-analysis of 26 samples ($n = 6024$), Pfennig and Hüsch (1994) found significantly negative correlations of all three burnout dimensions with self-esteem.

The explained average variance ranged from 14% (emotional exhaustion) to 10% (depersonalisation), and 9% (reduced personal accomplishment). Self-esteem refers to an individual's self-appraisal of competence and personal worth. Instead of a personality trait that may predispose individuals to burnout, poor self-esteem can also result from environmental factors (Tharenou, 1979), or perhaps even from burnout. The only longitudinal study that investigated the causal relationship of burnout and self-esteem found no conclusive evidence about their causal order (Rosse *et al.*, 1991).

Results concerning 'feeling types' and 'thinking types' are less clear cut. These personality types originate from the work of Carl Gustav Jung and are discussed in greater detail in Chapter 5 because they are supposed to be of special theoretical relevance. There are indications that individuals who prefer feeling to thinking are more prone to burnout, especially to depersonalisation (Garden, 1988; 1989). Unfortunately, these personality types seem to be poorly operationalised so that further research is necessary.

'Type-A behaviour' seems to be positively related to burnout, especially to emotional exhaustion (e.g. Burke, 1985; Nowack, 1986). Type-A individuals tend to be competitive, prefer a striving, time-pressured lifestyle, and show an excessive need for control. Research suggests that Type-A behaviour is strongly related to stress-induced illnesses, especially coronary heart disease (Miller *et al.*, 1996).

The 'Big Five' and burnout

It is of particular importance to see how those personality traits that have been summarised as the main underlying dimensions of personality, commonly referred to as the 'Big Five', are related to burnout. The popular Five-Factor Model of personality (McCrae and John, 1992) includes: (1) neuroticism; (2) extroversion; (3) openness to experience; (4) agreeableness; (5) conscientiousness. Numerous studies show that neuroticism and introversion are especially related to negative affect, poor well-being, somatic complaints, and poor personal adjustment (Clark *et al.*, 1994). *Neuroticism* includes trait anxiety, hostility, depression, self-consciousness and vulnerability. Neurotic individuals are emotionally unstable and prone to psychological distress. *Extroversion* comprises self-confidence, dominance, activity, and excitement-seeking. Extroverts show positive emotions, a higher intensity of interpersonal interactions and a higher need for stimulation.

Two studies are particularly well-suited to illustrate the relationship of burnout and the 'Big Five' personality factors because they used random samples that had response rates well over 80%. The first study not only found that neuroticism was positively related to emotional exhaustion (17% shared variance), but also that it moderated the effects of daily hassles: when exposed to such minor stressors 'neurotic' highway patrol officers reported higher levels of exhaustion than did their less 'neurotic' colleagues (Hills and Norvell, 1991). The second study, among physicians, confirmed the positive association between neuroticism and emotional exhaustion (31% shared variance) and additionally showed that neuroticism was significantly related to depersonalisation and reduced personal accomplishment (16% and 12% of shared variance, respectively) (Deary *et al.*, 1996).

Because Deary *et al.* (1996) did not simultaneously analyse all the 'Big Five' factors in order to evaluate their relative contribution to the variance shared with burnout, we performed a secondary analysis based on their data. It appears that emotional exhaustion is positively related to neuroticism and openness (shared variance 33%), depersonalisation is positively related to neuroticism and negatively to agreeableness (shared variance 20%), and personal accomplishment is significantly related to four personality factors: neuroticism, extroversion, openness, and conscientiousness (shared variance 25%). Based on a small sample, Piedmont (1993) also found rather strong correlations between the 'Big Five' personality dimensions and burnout. They predicted that burnout would occur 7 months later even after controlling for situational factors. Longitudinally, neuroticism and agreeableness at time one explained 42% of emotional exhaustion at time two. Cross-sectionally, neuroticism shared 25% of the variance with this burnout dimension. Finally, a meta-analytic study summarising 12 studies on burnout and anxiety, which is part of the 'Big Five' factor neuroticism, showed that this trait correlates most highly with emotional exhaustion (shared variance 23%), followed by depersonalisation (17%) and reduced personal accomplishment (12%) (Pfennig and Hüsch, 1994).

Thus it seems that neuroticism acts as a vulnerability factor that predisposes employees to experience burnout. However, it cannot be ruled out that the consistently observed positive relation between neuroticism and emotional exhaustion is due to an artefact resulting from overlapping items. In addition, negative affectivity, i.e. the general tendency to report things negatively, might play a role as a third variable that explains neuroticism as well as high levels of emotional exhaustion. The negative correlation of depersonalisation and agreeableness can be explained by the fact that agreeableness belongs to the interpersonal circumplex that is characterised by altruism, nurturance, caring, and social support as opposed to hostility, indifference to others, self-centredness, and non-compliance. Finally, the fact that reduced personal accomplishment is related to almost all 'Big Five' factors concurs with the view that this burnout dimension reflects the employee's personality rather than his or her reaction to a stressful situation (Shirom, 1989; Schaufeli and Van Dierendonck, 1993).

Work related attitudes

Already in the first theoretical accounts, high or unrealistic expectations have been seen as being responsible for the development of burnout (see Chapter 2). These expectations may refer to the organisation as well as to the professional's work with recipients. In particular, the latter has been equated with idealism. Idealism is assumed to cause excessive expenditure of energy and to make way for disillusionment and frustration as burnout develops, thus leading to the popular idealism → burnout hypothesis.

Although 20 studies exist on the relationship between particular expectations and burnout, their results are difficult to compare because different concepts are used such as omnipotence, irrational beliefs, idealism, unmet expectations, disillusionment, and outcome expectations. Furthermore, it is not always clear whether expectations refer to the organisation, to clients' progress, or to personal effectiveness. Perhaps the most adequate summary of findings is that results are inconclusive: about 50% of the studies found a positive correlation of expectations with burnout (e.g. Schwab *et al.*, 1986; Dyment, 1989), in 40% the correlation was not significant (e.g. Moore, 1984; Salerno, 1991), whereas three studies (Seabold, 1983; Enzmann and Kleiber, 1989; Kirk and Koeske, 1995) found a negative correlation that contradicted the idealism → burnout hypothesis. In addition, it seems that unrealistic expectations are more strongly related to reduced personal accomplishment than to both other burnout components.

Conceptually speaking it is crucial to differentiate between expectations of personal effectiveness and expectations of clients' progress. Both types of expectations seem to be related to different burnout components and seem to develop in different ways (Stevens and O'Neill, 1983): with growing experience, expectations of clients' progress may drop, but at the same time this can be counteracted by increased expectations of personal effectiveness. The only prospective study available showed a positive effect of optimistic and idealistic expectations, thus contradicting the idealism → burnout hypothesis: optimistic and idealistically motivated social workers reported lower levels of burnout at the 1-year follow-up compared with their more pessimistic and less idealistic fellows (Kirk and Koeske, 1995). So perhaps those who hold high expectations and show more enthusiasm are more efficient at motivating their clients and thus are more likely to be successful, which counteracts feelings of burnout. By contrast, if one's expectations are frustrated by adverse conditions, feelings of exhaustion and diminished accomplishment may develop. This is probably the reason why Reilly (1994) found the 'paradoxical result' that the commitment of nurses to their profession was negatively related to burnout, although at the same time it increased the effects of stressors on emotional exhaustion.

These inconsistent results point to the need for theory. High or unrealistic expectations are not clearly defined and their relationship to omnipotence, optimism, idealism or professional commitment is obscure. A clear conceptualisation of different kinds of beliefs and expectations is needed, including their role within a theoretical model of work motivation. For example, theoretically speaking, the conceptual distinction between self-efficacy expectations and outcome expectancies (Bandura, 1997; Leiter, 1992a) is superior to the fuzzy differentiation between expectations of personal effectiveness and expectations of clients' progress (Stevens and O'Neill, 1983).

Work and organisational characteristics

Work and organisational characteristics are divided into four groups (see Table 4.2): job-related stressors, client-related stressors, social support, and factors that determine self-regulation of work activities. The first two groups can be considered job demands, whereas the latter two are resources. After a brief look at the relationship of individual work and organisational characteristics to burnout (sub-section 1), we will take a closer look at the relative importance of these characteristics for each of the three burnout dimensions (sub-section 2). Finally, we contrast job-related stressors with client-related stressors in order to see which is most important for the development of burnout (sub-section 3). As noted before, longitudinal studies on the relationship between job demands and burnout are excluded from sub-section 1 and are considered separately in Section 4.4.

1. Individual work and organisational characteristics

Workload and time pressure explain about 25–50% of the variance of burnout, especially of emotional exhaustion. A meta-analysis of Lee and Ashforth (1996) showed that workload and time pressure share on average 42% (six studies) and 25% (five studies) of variance with emotional exhaustion, respectively. Relationships are much weaker with both other MBI-dimensions. The high correlation with workload must be qualified, however, because this stressor is often operationalised in terms of experienced strain so that considerable overlap in item content exists, especially with emotional exhaustion (see Section 4.1).

Role conflict and role ambiguity are moderately to highly correlated with burnout. Role conflicts occur when conflicting demands at the job have to be met. For instance, correctional officers are expected to facilitate the delinquent's rehabilitation (educational role) as well as to guard them (disciplinary role). Role ambiguity occurs when no adequate information is available to do the job well. For instance, nurses may be deprived of the doctor's essential medical knowledge. Whereas role conflict results in conflicting goals and behaviours, role ambiguity obstructs the development of goals that direct work behaviour. According to a meta-analytic study by Pfennig and Hüsch (1994), role conflict (49 studies) shares 24% of variance with emotional exhaustion, 13% with depersonalisation, and only 2% with personal accomplishment. The percentages for role ambiguity (38 studies) are 14%, 8%, and 10%, respectively.

Other job demands such as the number of hours worked per week, the amount of direct client contact, caseload, and the severity of clients' problems, are only studied occasionally. Generally, correlations with burnout are lower but nevertheless in the expected directions: employees experience more burnout when they work more hours per week, interact frequently with recipients, have high caseloads, and have to deal with severe client problems (e.g. Maslach and Jackson, 1984b; Gibson *et al.*, 1989; see also below).

Although correlations between support and burnout are less strong than for job demands, clear evidence exists for a positive relationship between lack of social support and burnout. A lack of social support from supervisors is especially related to burnout. On average, support from supervisors explains 14% of the variance of emotional exhaustion, 6% of depersonalisation, and 2% of personal accomplishment (13 studies; Lee and Ashforth, 1996). These results, however, could not be replicated in longitudinal studies on social support and burnout (Dignam, 1986; Gusy, 1995). For social support from co-workers the amounts of variance are 5%, 5%, and 2%, respectively (14 studies; Lee and Ashforth, 1996). Independently from a direct effect on burnout, social support might buffer the effects of stressors such that employees who receive more support are better able to cope with their job demands. To date, however, studies have either failed to show the existence of a buffer effect of social support (Graham, 1993) or the results are equivocal (Himle *et al.*, 1991; Miller, 1991).

Lack of feedback is positively related to all three burnout dimensions. Although there are only a few studies available, their results are quite consistent: a meta-analysis of six studies showed that lack of feedback explains 18% of the variance of emotional exhaustion, 12% of depersonalisation, and 9% of reduced personal accomplishment (Pfennig and Hüsch, 1994).

Similarly, participation in decision making and autonomy are consistently negatively related to burnout, although the latter relationship is much weaker. A meta-analysis of six studies revealed that participation in decision making shares 10% of the variance of emotional exhaustion, 3% of depersonalisation, and 9% of reduced personal accomplishment, whereas the percentages for autonomy (11 studies) are 2%, 2%, and less than 1%, respectively (Lee and Ashforth, 1996). However, at least six studies that show higher correlations of autonomy with burnout, especially with reduced personal accomplishment, are not included in this meta-analysis (Enzmann and Kleiber, 1989; LeCroy and Rank, 1986; Kelloway and Barling, 1991; Melchior *et al.*, 1997; Poulin and Walter, 1993a; Wallace and Brinkeroff, 1991). In these studies percentages of explained variance vary from 3% to 18%. Perhaps differences in study results are due to different operationalisations of autonomy.

2. Comparison of burnout dimensions

It is striking that reduced personal accomplishment correlates least with job demands (with the exception of role ambiguity – see later). Workload and time pressure in particular are virtually unrelated to personal accomplishment, whereas most studies found that these two stressors were strongly related to emotional exhaustion.

In order to compare the pattern of correlations of the three burnout dimensions with various job demands more systematically we analysed 27 recently published studies. Eleven studies (e.g. Himle and Jayaratne, 1990; Starnaman and Miller, 1992) showed small differences with respect to the three burnout dimensions, whereas in 15 studies (e.g. Jackson *et al.*, 1987; Huberty and Huebner, 1988) job demands clearly correlate least with personal accomplishment and most strongly with emotional exhaustion. Only in one study was the reverse pattern found (Siefert *et al.*, 1991). In this study, however, emotional exhaustion was measured with a single item which increases measurement error and therefore the likelihood of non-significant results. Lee and Ashforth (1996) arrived at a similar result in their meta-analysis of 50 studies (that partly overlap with the 27 studies mentioned above): job stressors correlate least with personal accomplishment.

In view of these results, it is quite remarkable that personal accomplishment shows a comparatively strong association with a lack of role ambiguity, feedback, participation in decision making and, depending on its operationalisation, autonomy.

These factors have in common that they foster self-regulated activity at work and are therefore also referred to as resources (Leiter, 1993). Evidently, such resources are likely to influence feelings of professional self-efficacy and personal accomplishment.[4] This is in agreement with the observation that personal accomplishment is relatively strongly related to confronting coping behaviour as well as to self-efficacy (Friesen and Sarros, 1989; Gil-Monte et al., 1993; Wallace and Brinkeroff, 1991). It is likely that working conditions or resources that foster self-regulated activity at work simultaneously facilitate confronting coping behaviour and promote a sense of self-efficacy and personal accomplishment.

3. Job-related versus client-related stressors

Two types of job demands or stressors are especially important for the theory of burnout and for the uniqueness of the construct: job-related stressors and client-related stressors. The former (e.g. work overload and time pressure), in particular, are encountered in every occupation, whereas the latter (e.g. confrontation with death and dying) are characteristic of the helping professions. If it is true that burnout results mainly from emotional challenges and problems in the interaction with recipients, common job-related stressors should have minor effects as compared with client-related stressors.

We compared the results of 16 studies and found that, overall and contrary to expectations, common job-related stressors such as workload, time pressure, or role conflicts correlate more highly with burnout than client-related stressors such as interaction with difficult clients, problems in interacting with clients, frequency of contact with chronically or terminally ill patients, or confrontation with death and dying. More specifically, in nine studies common job related stressors correlated clearly more highly with burnout than client-related stressors (e.g. Cordes et al., 1997; Melchior et al., 1997), in four studies both types of stressors were similarly related to burnout (e.g. Hamann et al., 1987; Pierce and Molloy, 1990), whereas only three studies confirmed the special importance of client-related stressors in relation to burnout – contact with terminally ill patients and conflicts in interactions (Dames, 1983; Leiter, 1988; Sarros and Friesen, 1987). The latter result must be qualified, however, because conflicts in interactions were not exclusively found in helper-client relationships but similarly in relationships with superiors, colleagues, or subordinates (Leiter, 1988; Sarros and Friesen, 1987).

Hence, it seems that, on empirical grounds, the assertion that burnout is particularly related to emotionally charged interactions with clients has to be refuted. Next, we consider two client-related stressors in somewhat greater detail because of their particular significance for burnout as being a syndrome characteristic of the human service professions: caseload and work with 'difficult' clients.

When reviewing the literature on caseload and burnout, Koeske and Koeske (1989) found contradictory results that could be explained by distinguishing between quantity and quality. That is, the sheer number of client contacts is not a good indicator of experienced client-related stress, because a heavy caseload simultaneously implies heavy workload and time pressure. By contrast, the type of client (e.g. terminally ill or dying patients), problems in interacting with clients, and the relative amount of negative as opposed to positive client contacts are more appropriate as indicators of client-related stress. This is illustrated, for instance, by the study of Dignam *et al.* (1986), who found a modest positive correlation between burnout (emotional exhaustion and depersonalisation) and negatively judged client contact, whereas the correlation between burnout and positively judged client contact was weakly negative.

Six studies are available that include stressors that are related to working with 'difficult' recipients such as 'confrontation with death and dying' and 'work with severely ill patients'. The study of Dames (1983) is the only one in which nurses who work with terminally ill patients showed significantly higher burnout levels than nurses in other settings, such as intensive care units. By contrast, Hare *et al.*, (1988), found no relationship between burnout and working with terminally ill patients. There was, however, a positive correlation between fear of dying and depersonalisation. Mallett *et al.* (1991) compared correlations between fear of death (related to oneself and to others) and emotional exhaustion and depersonalisation in samples of nurses in intensive care units and nurses in hospice care. In both settings, fear of dying was weakly correlated with emotional exhaustion and depersonalisation. Moreover, nurses in hospice care were less burned out than nurses in intensive care units. Finally, all nurses stated that the lack of staffing and the fact that they had to work with insufficiently qualified staff were the most stressful factors in their jobs. Obviously, working with 'difficult' patients is the least disturbing part of the nurses' job and other non patient-related demands are experienced as more stressful. In a similar vein, McIntosh (1991) found a positive and moderately strong correlation between nurses' emotional exhaustion and workload (i.e. working speed and physical demands). However, no significant relationship was found for confrontation with death and dying. Kleiber *et al.* (1995) also found that confrontation with death and dying was a minor stressor that barely predicted burnout 1 year later.

These results concur with a study of hospital doctors that revealed that time pressure was experienced as the 'number one stressor' (76–81% of answers), whereas contact with severely ill patients was mentioned as stressful only by 54% (Pröll and Streich, 1984). It is likely that those who are frequently confronted with 'difficult' patients develop adaptive mechanisms that prevent negative long-term effects such as burnout.

4.3 CONCOMITANTS AND POSSIBLE CONSEQUENCES OF BURNOUT

Table 4.3 shows the possible effects or concomitants of burnout which are classified into consequences for the individual, effects on work orientation and attitudes, and consequences for the organisation. Compared with studies on possible causes of burnout, there are only a few studies on the impact of the syndrome. What is more, many studies do not consider the variables that are listed in Table 4.3 as possible consequences. Instead, they are either studied as concurrent effects of stressful

Table 4.3   Correlates of burnout: concomitants and possible consequences

individual level

depression	+	+	+
psychosomatic complaints	+	+	+
health problems		(+)	
substance use		+	
spillover to private life		(+)	

work orientation and attitudes

job satisfaction	−	−	−
organisational commitment	−	−	
intention to quit	+	+	

organisational level

absenteeism and sick-leave	+
job turnover	+
performance and quality of services	(−)

Note: The number of plus or minus signs denotes the size and the direction of the relationship with burnout.

working conditions, or the purpose of the study is to illuminate the discriminant validity of the burnout construct. The latter applies especially to depression, psychosomatic complaints, and job satisfaction.

Individual level concomitants and consequences

Depression is a likely consequence of burnout and belongs to the most extensively studied correlates on the individual level, particularly with regard to its discriminant validity *vis-à-vis* burnout (see also Chapters 2 and 3). We found 12 studies (including six studies reviewed by Glass and McKnight, 1996) that report on correlations between depression measures and separate MBI-dimensions (e.g. Leiter and Durup, 1994; Dion and Tessier, 1994). Emotional exhaustion is most strongly related to depression: both concepts share 12–38% of their variance (weighted population average 26%). Depersonalisation shares 2–29% of variance with depression (average 13%), and personal accomplishment 3–20% (average 9%). Only one study (McKnight and Glass, 1995) explicitly tries to answer the question whether depression may be a cause or a consequence of burnout. Unfortunately, the authors could not find support for either causal relationship: it seems that burnout can be regarded as a consequence as well as a cause of depression. There are several explanations for the substantial correlation between depression and burnout, especially with emotional exhaustion. First, burnout and depression simply share common symptoms such as low energy, poor work motivation, and negative attitudes. Second, neuroticism, being a fundamental personality trait, may underlie depression as well as emotional exhaustion. According to McCrae and John (1992), depression is part of the neurotic personality and, as we have seen before, emotional exhaustion is also related to neuroticism. Third, common external causes might exist: for instance, stressful working conditions may independently lead to burnout as well as to depression. Based on the currently available empirical results, it is not possible to decide which explanation is most appropriate.

Table 4.3 differentiates between psychosomatic complaints and health problems. The former refer to subjectively measured complaints which are rather difficult to verify objectively (e.g. neurasthenic symptoms, musculoskeletal problems, heart or circulatory disturbances, and gastrointestinal complaints), whereas the latter are based on objective diagnoses (e.g. myocardial infarction). Both, however, are considered as somatic stress reactions, i.e. negative outcomes of particular psychobiological mechanisms. Generally speaking, somatic stress reactions result from frequent and/ or prolonged psychophysiological arousal. Probably, they are neither cause nor consequence of burnout as burnout is not characterised by arousal that is evoked in stressful situations but by negative mood that is accompanied by motivational and attitudinal changes (see Chapter 2). Therefore, it is less likely that burnout itself leads to psychosomatic complaints or health problems.[5] Nor is it plausible that health problems lead to the motivational or attitudinal changes characteristic of burnout. Instead, psychosomatic complaints and health problems are most likely to be concomitants of burnout.

A few studies are available that show a consistent positive correlation between psychosomatic complaints and burnout, in particular with emotional exhaustion (e.g. Conner, 1982; Schaufeli and Van Dierendonck, 1993). Both constructs share between 20% and 46% of their variance, whereas the relationships with depersonalisation (between 6% and 21%) and personal accomplishment (between 3% and 18%) are much weaker. It is noteworthy, however, that those complaints that may be more easily verified by objective diagnoses correlate less strongly with burnout. For instance, a recent study showed that self-reported heart problems and circulatory disturbances share only 8% of their variance with emotional exhaustion, in contrast to neurasthenic symptoms (38% of variance) (Kleiber *et al.*, 1998). Possibly, subjectively reported psychosomatic complaints and burnout are likewise influenced by negative affectivity.

Unfortunately, we could find only one single study that investigated objectively diagnosed health indicators in relation to burnout as measured by the MBI. Hendrix *et al.* (1991) observed a small but significant relationship between emotional exhaustion and the frequency of self-reported cold or flu episodes (1% of explained variance). However, no relationship was found with objectively assessed cholesterol ratio, a risk factor for cardiovascular disease. By contrast, using a self-constructed exhaustion scale, partly overlapping with the emotional exhaustion items of the MBI, Shirom *et al.* (1997) found that levels of emotional exhaustion predicted changes in cholesterol levels in the next 2–3 years in males (but not in females), after controlling for various confounds such as age, weight, smoking intensity, and alcohol consumption.

As far as self-reported frequency of various illnesses is concerned, Corrigan *et al.* (1995) reported a shared variance with emotional exhaustion plus depersonalisation of 18%. However, after controlling for the effects of social support, this relationship dropped to less than 10%. A similar relationship between emotional exhaustion and a self-report measure of the frequency of serious illness (12% of shared variance) was found by Bhagat *et al.* (1995); depersonalisation was only marginally related (2%) and reduced personal accomplishment was unrelated to serious illness. Finally, Landsbergis (1988) found a significant positive relationship between self-reported symptoms of coronary heart disease and emotional exhaustion (3% of shared variance) and depersonalisation (4%). The relationship with reduced personal accomplishment was not significant (2%). Thus, convincing empirical support for the often claimed relationship between burnout and (objectively diagnosed) health problems is still lacking, although significant correlations with self-report measures were found.

Five studies could be tracked that report on correlations of substance use with burnout. Landsbergis (1988) observed no significant relationship between smoking and burnout. In a study of police officers, Burke (1994) found no relationship between coffee consumption, smoking, alcohol use, consumption of medicaments, or drug use with any of the burnout dimensions. One study found slightly more substance use (composite measure of alcohol, cigarettes, drugs) among women who scored more highly on depersonalisation (Nowack and Pentkowski, 1994). In a similar vein, Ogus *et al.* (1990) report a small but significant correlation between depersonalisation and the use of pain medication among male teachers (explained variance 3%). Controlling for socially desirable responding in a heterogeneous sample of human service employees, Kleiber *et al.* (1998) observed significant relationships between self-reported substance use and all three burnout dimensions: depersonalisation and emotional exhaustion were related to increased alcohol consumption and medication use, and reduced personal accomplishment correlated significantly with smoking. However, in neither case does the explained variance exceed 2%. Hence, it seems that substance use is weakly, if at all, related to depersonalisation, probably because this burnout dimension is linked with psychological withdrawal and palliative coping behaviour.

In burnout research, spillover to private life has almost exclusively been discussed in terms of marital satisfaction and family stress. These variables, however, were often not regarded as a consequence of burnout but as another predictor in addition to work stress (e.g. Hendrix *et al.*, 1991; Burke and Greenglass, 1994). Because the causal order is open to debate we will treat marital quality and family stress as concomitants of burnout instead of consequences. The study of Zedeck *et al.* (1988) is very interesting as they investigated the relationship of employee burnout with their spouses' perceptions of home and family spheres in great detail.[6] Spouses' ratings of employees' time with family, the employee losing his/her temper, and his/her interest in family matters were the most important variables of the family sphere that were related to burnout in the expected direction. Percentages of explained variance ranged from 2% to 7% for emotional exhaustion, from 1% to 4% for depersonalisation, and were 2% or less for reduced personal accomplishment. Thus, correlations of family life variables with burnout were higher for emotional exhaustion than for depersonalisation and least for reduced personal accomplishment. The remaining studies showed either no relationships between family problems or marital satisfaction and burnout, or correlations were rather low: 4% to 12% shared variance with emotional exhaustion, 4% to 6% with depersonalisation, and 2% to 4% with reduced personal accomplishment (e.g. Hendrix *et al.*, 1991; Leiter, 1992b). Furthermore, it seems that in general, family stresses or family hassles and burnout correlate somewhat more highly than marital satisfaction and burnout. In contrast to the previous cross-sectional studies, the only longitudinal study on spillover effects found a small positive effect of emotional exhaustion on marital satisfaction among female health care professionals (Leiter and Durup, 1996). The authors explain this counterintuitive finding by compensatory support of stressed employees from their spouses that may strengthen their marital relationship. Taken together, there is no conclusive evidence on negative spillover of burnout to private life.

Work orientation and attitudes

Job satisfaction, organisational commitment, and the intention to quit the job or leave the profession are subsumed under this heading. Although interesting from a conceptual point of view, we do not include withdrawal in our summary, because this construct is operationalised too heterogeneously in the few studies that are available: as behavioural withdrawal from work and clients (Enzmann, 1996); involvement and identification with the organisation (Dekker and Schaufeli, 1995); observed avoidance of client contacts (Lawson and O'Brien, 1994); and cognitive withdrawal (Van Gorp *et al.*, 1993). Job satisfaction belongs to the most extensively studied concomitants or consequences of burnout. Organisational commitment is conceptually related to the intention to quit because it refers to 'a strong desire to maintain membership in the organization' (Mowday *et al.*, 1979, p. 226). Nevertheless, it is important to distinguish between both constructs because the intention to quit the job can be considered a consequence of negative attitudes displayed in poor organisational commitment.

Recently, Lee and Ashforth (1996) conducted a meta-analysis of the relationships between burnout and job satisfaction, organisational commitment, and the intention to quit. We reanalysed their findings after including 15 additional studies. Table 4.4 shows the weighted mean correlations after correction for measurement unreliability.

Different operationalisations of job satisfaction exist, including (single-item) global measures, measures that tap intrinsic aspects such as challenge, meaningfulness, or stimulation, and measures that tap extrinsic aspects (e.g. satisfaction with salary and promotion). Job satisfaction correlates comparatively highly with all three burnout dimensions but most highly with depersonalisation (population weighted shared

Table 4.4   Meta-correlations of job satisfaction, organisational commitment, and intention to quit with MBI dimensions

		emotional exhaustion	depersonalisation	reduced personal accomplishment
job satisfaction	studies	25	20	19
	n	6516	4618	4757
	r	−.44	−.52	−.40
	shared variance	20%	27%	16%
organisational commitment	studies	9	8	9
	n	3956	3483	3956
	r	−.40	−.40	−.22
	shared variance	16%	16%	5%
intention to quit	studies	13	10	6
	n	3124	2457	1959
	r	.45	.35	.24
	shared variance	20%	12%	6%

Note: r = weighted population effect size (correlation) corrected for unreliability of measurement.

variance: 27%), followed by emotional exhaustion, and reduced personal accomplishment (20% and 16% shared variance, respectively). Besides, intrinsic satisfaction and burnout correlate somewhat more highly than extrinsic satisfaction and burnout. Since virtually all studies are cross-sectional, no causal order can be determined. A notable exception is the longitudinal study of Wolpin *et al.* (1991) who only found indirect effects of burnout on job satisfaction one year later. Thus their conclusion that burnout causes diminished job satisfaction instead of the other way around seems to be rather premature.

In most studies, organisational commitment is assessed with the Organisational Commitment Questionnaire (Mowday *et al.*, 1979) that measures the employee's belief in and acceptance of the goals or values of the organisation, the willingness to exert effort on behalf of the organisation, and the desire to maintain one's membership. Although less strongly than job satisfaction, organisational commitment consistently correlates negatively with emotional exhaustion and depersonalisation (shared variance 16%). The relationship with reduced personal accomplishment is clearly weaker (shared variance 5%). Similar results are found with respect to the intention to quit which shares 20% of variance with emotional exhaustion, 12% with depersonalisation, and 6% with reduced personal accomplishment.

In sum, the strength of the association of burnout with negative work orientations, attitudes, and intentions is impressive. Whether burnout also affects the employee's organisational behaviour, for example, whether the intention to quit actually results in job turnover remains to be seen and will be considered below.

Organisational level concomitants and consequences

At the organisational level, the most important concomitants and consequences of burnout are absenteeism, job turnover, and impaired performance. These are behaviours that can be observed or assessed objectively, for example, by using company records, or by using performance ratings from superiors. It should be expected that such objective measures correlate less strongly with burnout than self-report measures because they are neither influenced by common method variance nor by negative affectivity or socially desirable responding.

Because we could only obtain ten studies that included absenteeism we do not differentiate between various ways in which it has been operationalised. For instance, some studies use frequency of absent spells, whereas in other studies the number of absent days is employed. Sometimes general absenteeism or unspecified sickleave is assessed and sometimes sickleave due to mental or physical causes. Furthermore, self-reports or company records are used, whereas time intervals vary from 6 weeks to up to 1 year. Overall, across these different operationalisations the relationship with emotional exhaustion is most consistent (2% explained variance on average), followed by depersonalisation (1%) (e.g. Price and Spence, 1994). By contrast, reduced personal accomplishment is positively related to absenteeism only in three studies with an average explained variance of less than 1% (Kimmel, 1993; Parker and Kulik, 1995; Kleiber et al., 1998). Thus, despite the popular assumption that burnout causes absenteeism, the effect is rather small and is most related to emotional exhaustion.

Contrary to studies on turnover intentions, the literature on burnout and actual job turnover is rather scarce: we could only obtain four publications in which effects of separate MBI-dimensions on turnover rates are reported. Two studies found a positive effect of depersonalisation on job turnover two years later among nurses (Firth and Britton, 1989)[7] and among human service professionals (Kleiber et al., 1998), respectively. Two other studies found a similar significant effect of emotional exhaustion among teachers within one year (Jackson et al., 1986) and among general practitioners within 5 years (Sixma et al., 1998). In terms of shared variance, the significant effects were rather low, ranging between 1% and 5%. No effects were found with respect to reduced personal accomplishment. The fact that the relationship of turnover intentions with burnout (see Table 4.4) was much stronger than with actual turnover, suggests that a large percentage of burned out professionals stayed in their jobs involuntarily (see also Cherniss, 1992). This might have negative consequences for the individual employee as well as for the organisation (Jackson et al., 1986).

The relationship of burnout with performance is interesting from a theoretical as well as from a practical point of view. Conceptually speaking, the symptoms of burnout not only include the individual's dysphoric state but also his or her impaired performance. This is reflected in various definitions that were discussed in Chapter 2 (see Brill, 1984) as well as in the MBI dimension labelled personal accomplishment. The important theoretical question is: do those feelings of reduced accomplishment result in an actual reduction of performance, or do they reflect a negative self-evaluation that is merely another indicator of the individual's negative dysphoric state? If burnout is indeed accompanied by poor performance this would validate the burnout construct, especially the MBI personal accomplishment dimension. Also, from a practical point of view a significant relationship between burnout and poor performance is of great importance: it would mean that the organisation's quantity and quality of services would be severely impaired when the employees were burned-out.

It is important to distinguish between self-ratings of performance and objective measures or ratings by others such as co-workers or supervisors. Based on six studies that include over 2000 employees, we conclude that self-rated performance correlates weakly with burnout: on average, 5% of variance is shared with emotional exhaustion, 4% with depersonalisation, and 6% with reduced personal accomplishment (Keijsers et al., 1995; Lee and Ashforth, 1990; Parker and Kulik, 1995; Ursprung, 1986; Zedeck et al., 1988; Zwerts et al., 1995). Moreover, it is noteworthy that, on balance, correlations with personal accomplishment are not higher than correlations with the two other burnout dimensions.

With respect to other-rated or objectively assessed performance, the results are inconsistent and disappointing: three studies found no significant or even positive correlations with burnout (Keijsers et al., 1995; Lazaro et al., 1985; Randall and Scott, 1988), whereas four studies found the expected significant negative correlations, at least with some burnout dimensions (Bhagat et al., 1995; Hendrix et al., 1991; Nowack and Hanson, 1983; Parker and Kulik, 1995). However, on average the explained variance is less than 1%, irrespective of the dimension of burnout.

Two studies are especially interesting because of the unique way in which they assessed performance. In a study of a small sample of direct care staff who work with the developmentally disabled, Lawson and O'Brien (1994) assessed the quality of work performance over several weeks by employing a standardised activity rating scheme based on unobtrusive observations of a range of desirable and undesirable behaviours. Although the observations of staff activity confirmed the existence of burnout on a behavioural level (for example, avoidance of client contact), scores on the MBI were rather low. Only one of eleven activity measures (positive interaction with clients) correlated significantly with emotional exhaustion and none was related to depersonalisation or reduced personal accomplishment. In a large-scale study of burnout among nurses in intensive care units (ICUs), Keijsers et al. (1995) obtained an objective measure of ICU performance by calculating for each unit a standard mortality ratio, the ratio of actual versus predicted death rates adjusted for several patient characteristics. Contrary to expectations, they found a small but significant *positive* correlation of objective ICU performance with emotional exhaustion (explained variance 2%) and no relationship with depersonalisation or personal accomplishment. More detailed analyses showed that those nurses who were employed in objectively and subjectively well-performing ICUs, but who scored low on self-reported personal accomplishment, felt especially exhausted. A possible explanation is that nurses in well-performing ICUs actually exert themselves more and as a consequence they become more exhausted (see also Schaufeli *et al.*, 1995). An alternative explanation is that nurses in well-performing units have a higher standard of comparison and thus feel that they accomplish less.

At any rate, both studies show that, in contrast to the prevailing view, burnout is not necessarily linked to low levels of actual performance. It is especially noteworthy that this also holds for feelings of personal accomplishment.

4.4 LONGITUDINAL EFFECTS OF JOB DEMANDS

In Section 4.1, we have seen that, compared with specific client-related stressors, common quantitative and qualitative job demands are more strongly related to burnout, especially emotional exhaustion. However, so far we have mainly discussed cross-sectional evidence. In this section we will review longitudinal studies on the effect of job demands on burnout. We only include longitudinal studies that measured job demands *and* burnout at least twice. Furthermore, we only included those studies that controlled for levels of burnout at the first time of measurement. Principally, this can be achieved either by regression techniques that are based on correlations between time one and time two measures of burnout or by computing change scores, that is the differences between time one and time two measures of burnout.

Empirical results

Because the eight studies that satisfy the above mentioned requirement are too heterogeneous in design, operationalisation of variables and analytic strategy, we will discuss their results separately.[8]

Among prison personnel, burnout at the second point of measurement was predicted exclusively by burnout at the first point of measurement (Dignam 1986; see also Dignam and West, 1988). A composite measure of emotional exhaustion and depersonalisation was used. Longitudinal effects of work stress (a composite of role ambiguity, role conflict, and workload) and social support from co-workers and supervisors were not significant, although in cross-sectional studies they had proved to be significant predictors of burnout. As a possible explanation of this disappointing result, the authors point to the comparatively short time interval of 3 months.

In the longitudinal study of Shirom and Oliver (1986) in which teachers were surveyed twice with an interval of 10 months, burnout was also not predicted by stressors such as heterogeneous classes, necessity to punish pupils, and conflicts between work and private life. The longitudinal effects were ambiguous and with respect to two stressors (physical working conditions and extra-curricular teaching demands) burnout even seemed to be a predictor rather than a consequence.

Based on a small sample of helping professionals, Wade *et al.* (1986) investigated longitudinal effects of the work environment on burnout over a 1 year period.
As predictors of burnout at the first time of measurement they included involvement, peer cohesion, supervisor support, autonomy, task orientation, work pressure,
clarity, control by rules and pressures of management, innovation, and physical
comfort. They created two groups: one with employees whose burnout scores had
increased, and one that included those with decreased scores. Small but significant
effects were found only for peer cohesion, which was related to a decrease of burnout, and experienced control through management, which was related to an increase
of burnout.

Severe problems in predicting burnout appeared in the longitudinal study of Lee
and Ashforth (1993) who investigated a sample of superiors and managers with a
time interval of 8 months. The longitudinal effects of role stress and social support
on emotional exhaustion were significant (after controlling for stability). However,
contrary to expectations the relationship was negative: role stress and lack of support caused lower levels of exhaustion instead of higher levels. This was all the
more remarkable since the cross-sectional correlations at both points in time were
positive. The authors explain the contradicting results by the high stability of the
stressors as well as of burnout. This would preclude a valid interpretation of the
longitudinal effects (see also Section 4.4).

Poulin and Walter (1993b) investigated the extent to which an increase or decrease
in emotional exhaustion across a period of 1 year is related to initial levels of several
job demands and resources (job stress, supervisory support, job autonomy, organisational resources, severity of client problems, time spent with clients) among social
workers. Whereas all initial job demands and resources (except severity of client
problems) correlated significantly and in the expected directions with changes of
emotional exhaustion, most correlations disappeared when change-scores of demands
and resources were used. This means that changes in emotional exhaustion do not
correspond with similar changes in demands or resources. In addition, multivariate
analyses showed that job stress was the most important predictor of change (28% of
variance). However, the results must be interpreted with caution because of the use
of extreme groups and the imprecise description of the analytic strategy employed.

Burke and Greenglass (1995) studied teachers with a time interval of 1 year.
They found that, contrary to expectations, 'sources of stress' (a composite of doubts
about competence, problems with clients, bureaucratic interference, lack of stimulation and fulfilment, lack of collegiality) at time one was negatively related to burnout (a composite of the three MBI dimensions) at time two when the stability of all
variables was taken into account. The more stress, the *lower* the burnout levels were
1 year later. Again, as in Lee and Ashforth's (1993) study, cross-sectional correlations were positive. However, in contrast to these authors, Burke and Greenglass
avoid discussing this outcome as being a negative result.

In a somewhat similar vein, Enzmann (1996) also reports contradictory results.
He studied a heterogeneous sample of human service professionals at three points
in time with intervals of 11 months. Although significant and expected longitudinal
effects of time pressure, lack of decision latitude and autonomy, and problems in

client interactions on emotional exhaustion were found, they could only be observed among the less experienced. Moreover, these findings were observed only across the second time interval and could not be replicated with data from the first interval. Possibly, this was caused by a major event that took place during the first time interval and that may have changed the lives of many respondents living in Germany, where the study was carried out: the fall of the Berlin Wall.

The longitudinal study of Leiter and Durup (1996) among health care professionals that used a time interval of 3 months also rendered unexpected findings. They found, after controlling for stability, that emotional exhaustion predicted work overload and supervisor support, instead of the other way round. Although the effect of work overload on depersonalisation was as expected, it appeared to be rather small. Besides, no longitudinal effects were found for personal accomplishment.

Conclusions

The eight most sophisticated longitudinal studies that investigated effects of work demands on burnout by controlling for the degree of burnout at the first point of measurement could not reproduce the results found in cross-sectional studies. Either longitudinal effects were very small or not significant (Dignam, 1986; Enzmann, 1996; Leiter and Durup, 1996), or, contrary to expectations and despite positive cross-sectional correlations, demands seemed to be associated negatively with burnout (Shirom and Oliver, 1986; Lee and Ashforth, 1993; Burke and Greenglass, 1995). A reverse causal direction was even found, suggesting that emotional exhaustion leads to high workload rather than the other way around (Shirom and Oliver, 1986; Leiter and Durup, 1996). It is noteworthy that both the studies that used simple change-scores of burnout (time one minus time two scores), instead of a regression approach, seem to yield results that are more in line with expectations (Wade *et al.*, 1986; Poulin and Walter, 1993b).

How can these rather disappointing longitudinal results be explained? Perhaps burnout is truly unrelated to job demands. However, in Chapter 5 we will see that this is not very likely from a theoretical point of view. Hence, methodological explanations seem to be more plausible. First, a general tendency of respondents to report negatively on all self-report measures involved, burnout and job demands, might exist resulting in spuriously high cross-sectional correlations that are not likely to be reproduced longitudinally. This artefact, for which a particular personality characteristic (negative affectivity) may be held responsible, would explain the striking difference between cross-sectional and longitudinal correlations (Dignam, 1986; Lee and Ashforth, 1993; Burke and Greenglass; 1995; Shirom and Oliver, 1986). Secondly, the regression approach to control for the initial burnout status may be an inappropriate method to study predictors of change, especially if the stability of scores is high, which seems to be the case for burnout as is shown in the next section. This would explain why studies that used the regression approach in particular yielded negative results. Recently, promising alternative methods for the study of change have been developed, such as growth curve modelling (see

Gottman, 1995), that could be used in future burnout research. In sum, it is likely that the unexpected and disappointing longitudinal results are due to methodological problems and may not be interpreted as evidence against the causal job demands → burnout sequence.

4.5 STABILITY AND CHANGE

Do burnout scores wax and wane with time or do they remain rather stable? Table 4.5 shows 15 studies that assessed the stability in percentages of shared variance of the MBI dimensions for different time intervals.

Table 4.5 Stability of burnout (MBI) as shown in longitudinal studies

study	n	stability EE in per cent		stability DP in per cent		stability PA in per cent		interval in months
Dignam, 1986[a]	171	67	(83)	61	(93)		–	3
Leiter and Durup, 1996	151	56	(70)	41	(72)	38	(61)	3
Capel, 1991[b]	232	51	(60)[1]	43	(57)[2]	36	(45)[3]	4
		46	(54)[1]	44	(59)[2]	39	(49)[3]	5
		64	(75)[1]	29	(39)[2]	62	(78)[3]	9
Leiter, 1990	122	35	(48)	25	(50)	40	(66)	6
Piedmont, 1993	29	47	(55)[1]	59	(79)[2]	62	(78)[3]	7
Corrigan et al., 1994	35	53	(65)	55	(–)	40	(80)	8
Lee and Ashforth, 1993	143	55	(63)	52	(80)	42	(61)	8
Van Dierendonck et al., (in press)	245	35	(49)	41	(90)	44	(76)	12
Jackson et al., 1986	248	37	(52)	31	(56)	35	(58)	12
Poulin and Walter, 1993b	879	44	(53)		–		–	12
Wade et al., 1986	46	38	(45)[1]	12	(16)[2]	20	(25)[3]	12
Enzmann, 1996	192	–	(36)	–	(62)	–	(56)	22
McKnight and Glass, 1995	100	24	(28)[1]	41	(55)[2]	23	(29)[3]	24
Rosse et al., 1991	335	41	(56)	29	(47)	27	(48)	19–43
Bakker et al., 1997	209	–	(45)	23	(52)	46	(86)	60

Notes: Values in brackets: stability corrected for measurement unreliability (computations by the authors); EE: emotional exhaustion, DP: depersonalisation, PA: personal accomplishment;
[a] modified EE-scale (6 items);
[b] sum of frequency and intensity dimension; 9 months interval = interval 1 + 2;
[1] estimated reliability: $\alpha = .85$;
[2] estimated reliability: $\alpha = .75$;
[3] estimated reliability: $\alpha = .80$.

As far as the uncorrected stability values of emotional exhaustion are concerned, between 24% and 67% of the variance of the second measurement is explained by the first measurement. After correction for unreliability of measurement in most cases, more than 50% of the variance is explained. Regarding the other burnout dimensions, stability values are similarly high: 12–61% of variance for depersonalisation, and 20–62% for personal accomplishment (corrected 16–93% and 25–86%, respectively). Remarkably, the length of the time interval seems to be unrelated to the stability of burnout: correlations across a 6 month interval (Leiter, 1990) are similar to those across a 2 year period (Rosse et al., 1991) or even a 5 year interval (Bakker et al., 1997). The weighted mean stabilities estimated for a 1 year interval are 45% (corrected 56%), 41% (corrected 69%) and 40% (corrected 61%) for emotional exhaustion, depersonalisation, and personal accomplishment, respectively.[9] For a construct that is supposed to assess a state that is influenced by current situational characteristics, these values are rather high. Personality traits such as neuroticism and extroversion generally show stability values of 35% or more for a time interval of several decades (Finn, 1986). For example, an analysis of the data presented in Conley (1984) renders quite similar stability coefficients of neuroticism (40%, or 59% after correction for unreliablity of measurement) for an interval of 1 year.

Capel (1991) found a counterintuitive result that points to an important issue. For teachers, stability of emotional exhaustion and reduced personal accomplishment appeared to be higher across a longer time interval of 9 months compared with shorter sub-intervals of 4 and 5 months, respectively. Most likely this was due to the specific sample under study: teachers experience stress periodically depending on certain events which are season-bound (e.g. holidays, exams; see also Westman and Eden, 1997). Accordingly, levels of burnout are higher at particular times during the year. Capel's study shows that longitudinal results can be strongly influenced by the chosen time interval and by periodic events, and must be generalised with utmost care.

Finally, it is important to raise an issue that generally causes much confusion: high correlations of burnout scores across time do not imply that the mean burnout levels do not change! Stability coefficients based on correlations between measures are indicators of predictability (see Chapter 3) and reflect the stability of the rank order of the individuals over time. For example, although the stability coefficients reported by Enzmann (1996) are rather high (see Table 4.5), they concur with a significant *increase* of the mean scores of all three burnout dimensions over almost 2 years. By contrast, the high stability coefficients of Bakker et al. (1997) (see Table 4.5) concur with a significant *decrease* of the mean scores of all three burnout dimensions within 5 years. Thus, the increase or decrease of burnout levels is basically independent of their stability. This shows that stability and change are different concepts that must not be confused.

What can be concluded about the stability of burnout? First, the relatively high stability coefficients displayed in Table 4.5 indicate that burnout is a chronic problem rather than a transient state. Second, these coefficients suggest that burnout might be determined by personality traits such as hardiness or neuroticism, or by unchanged situational characteristics such as high job demands. Finally, since burn-out is stable across time the possibility that particular causes of burnout may be uncovered by using standard methods is reduced. Put simply, because of these high correlations across time there is little room left for additional factors to exert their influence. In fact, the likelihood of successfully uncovering causal agents of burnout by using cross-lagged regression models is inversely proportional to its stability: the more stable the construct, the less likely its antecedents are to be identified.

4.6 SUMMARY

Our review of the empirical literature can be summarised in five points:

■ Self-report measures correlate much more highly with burnout than data based on records, observations, or assessments by others such as supervisors or co-workers. This may explain, at least partly, why organisational consequences (Table 4.3) that are not exclusively based on self-reports are only weakly related to burnout.

■ As far as the three dimensions of the MBI are concerned, emotional exhaustion is most strongly related to possible causes, concomitants, and consequences of burnout. This applies especially to work-related stressors, neuroticism, depres-sion, and psychosomatic symptoms. Generally, correlations with depersonalisation are weaker. It is noteworthy, however, that this attitudinal dimension of burnout correlates most strongly with job satisfaction, which can also be regarded as a work-related attitude. On balance, personal accomplishment is least strongly related to potential correlates. The exceptions are the 'Big Five' factors of personality, confronting coping behaviour, job characteristics that facilitate the regulation of goal directed behaviour, and subjective performance. These results confirm the validity of personal accomplishment as a burnout dimension that reflects professional self-efficacy (Cherniss, 1993; Maslach and Leiter, 1997).

■ Comparing the relative importance of correlates of burnout, hardiness, neuroticism (anxiety), workload, and time pressure seem to be the most important (potential) causes, whereas depression, psychosomatic complaints, and job satisfaction are the most important concomitants or consequences. Quantitative demands such as time pressure and workload are clearly more strongly related to burnout than qualitative demands such as problems in interacting with clients or the con-frontation with death and dying. This contradicts the popular assertion that interaction with clients and the confrontation with their emotional needs is at the heart of burnout.

■ The weakest or most inconsistent relationships of potential correlates with burn-out were found with respect to the level of education, high or unrealistic expecta-tions, objective health problems, spillover to private life, and objectively assessed performance. It is noteworthy that personal accomplishment is weakly related to objective performance measures at the organisational level.

■ When controlling for the initial status of burnout, longitudinal studies could not replicate the findings of cross-sectional studies with respect to the effects of job demands on burnout. Instead, the most important predictor of burnout was burnout, measured at an earlier point in time. However, although the stability coefficients of burnout are high, significant changes of mean scores over time could be observed.

Taken together, the foregoing results lead to one overall conclusion: a revision of burnout theory is badly needed. Therefore, in the next chapter we discuss the theoretical approaches to burnout in greater detail.

Notes

1   Psychological Abstracts, Medline, Resources in Education/ERIC, Sociological Abstracts, Dissertation Abstracts International, National Union Catalog (Library of Congress).
2   National Library of Australia, 1981; Miller and Kobelski, 1982; Gillespie, 1983; Riggar and Beardsley, 1983; Riggar, 1985; Pletcher, 1987; Read, 1987; Kleiber and Enzmann, 1990; Lubin et al., 1992.
3   Although anxiety has quite often been studied as a separate personality characteristic in relation to burnout, we prefer to discuss it in the context of neuroticism under which heading it is subsumed in the 'Big Five' model of personality.
4   The comparatively high correlation between role-ambiguity and reduced personal accomplishment might also be a methodological artefact: both are the only stress- and burnout-scales with positively phrased items (role-clarity and personal accomplishment). Thus, the correlation could alternatively be explained by response tendencies.
5   It should be noted, however, that vital exhaustion, partly overlapping with symptoms of emotional exhaustion, has been observed as a precursor of myocardial infarction (Appels and Schouten, 1991; Kop, 1994; see also Melamed et al., 1992).
6   Another noteworthy study investigated the crossover of strains among married couples (Westman and Etzion, 1995). Because burnout was not measured with the MBI in this study it will not be discussed here.
7   By applying a two-tailed test, however, this result would not have reached the level of significance.
8   We restrict ourselves to the results and conclusions presented by the authors, although in some cases (Burke and Greenglass, 1995; Lee and Ashforth, 1993; Shirom and Oliver, 1986) a re-analysis of the data may yield different interpretations.
9   By employing weighted linear regression analyses of Fisher's Z-scores on time (compare Hedges and Olkin, 1985, p. 241).

CHAPTER FIVE

How to explain burnout?

Theoretical approaches

Practitioners such as consultants and interventionists, rather than academic scholars, were among the first to adopt the term 'burnout', a lay-term after all. Because of its popular background the scientific community was quite reluctant to accept its legitimacy as a theoretical construct. For instance, at the beginning of the 1980s burnout was still not included in most handbooks on job stress although by that time approximately 500 publications on that subject had appeared! Obviously, such a negative and sometimes even devastating scholarly climate does not encourage the development of theoretical approaches to burnout.

Initially, most theorising was rather speculative and eclectic, borrowing concepts from various psychological theories. Recently, more specific theoretical approaches have been developed and a growing number are supported by empirical findings (see Schaufeli *et al.*, 1993b). Instead of merely applying theoretical notions from well-established theories these more recent approaches try to understand the phenomenon. Despite the efforts of the past decades, a comprehensive theoretical framework for burnout is still lacking. As a matter of fact, a generally accepted theory or an all-embracing explanatory model of burnout will always remain illusory in view of the huge complexity of the phenomenon. Accordingly, we have to live with different coexisting theoretical perspectives.

This chapter distinguishes four sets of theoretical approaches to burnout. *Individual* approaches emphasise the role of factors and processes within the person, whereas *interpersonal* approaches focus on demanding relationships with others at work. *Organisational* approaches emphasise the relevance of the organisational context, whilst *societal* approaches focus on the broader social and cultural dimensions of burnout. These four approaches are not mutually exclusive. Rather, they differ in the extent to which they stress the importance of particular types of factors in the development of burnout. In the closing section, an attempt is made to integrate the common and most important aspects of burnout into a synthetic model.

5.1 INDIVIDUAL APPROACHES

Most of the eight individual approaches to be discussed are speculative since they are mostly not supported by empirical evidence. Typically, these approaches attempt to analyse burnout from general psychological perspectives, some of which are well-known (psychodynamic theory, cognitive psychology, or learning theory), whereas others are somewhat less familiar (conservation of resource theory, action theory).

Burnout as a failure to retain one's idealised self-image

The work of Herbert Freudenberger, the founding father of the burnout syndrome, does not contain a coherent psychological explanation of burnout. As a practising psychiatrist, Freudenberger presents a wealth of case-material that illustrates his rather loosely formulated principles. According to Freudenberger and Richelson (1980, p. 22) burnout is limited to 'dynamic, charismatic, goal-oriented men and women or to determined idealists'. Therefore, it is called 'the disease of over-commitment' or 'the super-achiever sickness'. The burned-out cases who are portrayed in the best-selling 1980 book *Burnout – How to Beat the High Cost of Success* are, without exception, strongly dedicated and committed individuals. They all have unrealistic or high expectations and are characterised by burning ambitions.

According to Freudenberger and Richelson (1980), burnout develops when individuals firmly believe in their idealised images of themselves as charismatic, dynamic, inexhaustible, and super-competent persons. As a result they lose touch completely with their other more fallible 'real' selves. In vigorously trying to live up to their idealised self-images, burnout candidates typically use the wrong strategies, further depleting their emotional resources. These 'false cures' are summarised in the four Ds: disengagement, distancing, dulling, and deadness. Fortunately, Freudenberger and Richelson (1980) knew how to cure burnout: the magic word is closeness. Just as the four Ds are the allies of burnout, closeness is its foe. Before one can achieve closeness with others, one has to get in touch again with one's real authentic self. Once this is accomplished, the fire which is burning brightly inside is warming rather than consuming.

Implicitly, Freudenberger uses an energy depletion metaphor: the burned-out individual pursues too long, too frantically, a much too idealistic self-image which tends to drain all energy resources. Because wrong coping strategies are used, the process of energy depletion is intensified instead of stopped or slowed down. According to Freudenberger, burnout sets in, not because of a lack of competence to achieve one's goals, rather it results from not obtaining the expected rewards.

Burnout as progressive disillusionment

Based on their observations in burnout workshops they conducted with human services professionals, Edelwich and Brodsky (1980) formulated their popular four-stage model of progressive disillusionment. The basic tenet of their model is that the initial idealistic expectations of the individual are frustrated by everyday reality. They argue that noble aspirations are difficult to realise because of several built-in sources of frustration in the human services: lack of criteria for measuring accomplishment, low pay, poor career perspectives, inadequate institutional support, and low social status. Hence, a gradual process of burning out may set in characterised by the following four stages of progressive disillusionment:

- **Enthusiasm** Initially, the professional is full of energy, works hard and the job promises to be everything. This stage is characterised by enthusiastic idealism, high hopes, and unrealistic expectations. For instance, a burned-out social worker from a hospital detoxification unit remembers: 'I was going to be the Martin Luther King of the drunks, leading them to the promised land' (p. 44).

- **Stagnation** The high initial expectations are markedly reduced. For instance, a teacher found her expectations scaled down to 'keeping the students interested for 15 minutes at a time' (p. 45). Instead of overidentification with recipients, as in the previous phase, the emphasis is now on meeting one's own personal needs. Other issues like money (the pay-cheque) and working hours (nine to five) become important.

- **Frustration** Frustration develops because of increasing powerlessness. The idealistic expectations from the first stage are not fulfilled, nor are the other personal needs that characterised the previous stage satisfied. At this point, external factors such as low pay, lack of institutional support, and bad office politics play a crucial negative role. Typically, affective, cognitive, and physical symptoms are observed at this stage.

- **Apathy** The professional withdraws from the job, physically as well as mentally. Spells of absenteeism are observed and social contacts with others at work are avoided, whereas, on the psychological level, emotional detachment, cynicism and numbness occur.

By distinguishing four successive stages, Edelwich and Brodsky (1980) have contributed to the understanding of burnout as a dynamic process that gradually develops over time. The driving force behind this process is the discrepancy between one's own dreams and the real world:

> The seeds of burnout are contained in the assumption that the real world will be in harmony with one's dreams.   (p. 16)

Unfortunately, for many this proves not to be the case. Like most stage models, the model of progressive disillusionment has strong intuitive appeal. Yet, virtually no empirical evidence exists that confirms the model, except for the cases, vignettes and anecdotes that are included in Edelwich and Brodsky's book.

Burnout as a pattern of wrong expectations

Meier (1983) criticised the affective emphasis exemplified in the work of Freudenberger (1980) and Edelwich and Brodsky (1980) and proposed a cognitive-behavioural approach to burnout. According to this approach, burnout develops as a result of a pattern of wrong expectations that do not correspond with the actual work situation. Three kinds of expectations are distinguished:

- **Reinforcement expectations** that pertain to the value and meaning of certain work outcomes. For instance, a particular teacher prefers working with pupils who frequently ask questions during class, whereas a colleague finds a silent class of students more satisfying. When such personal reinforcement expectations are not met burnout is likely to occur.

- **Outcome expectations** that pertain to behaviours leading to certain work outcomes. For instance, based on her past experience, a teacher may believe that a particular exam is simply too difficult for the pupils. The teacher therefore expects that, whatever effort is put into teaching the students, most of them will fail anyway. This teacher might come to experience 'learned helplessness': since no apparent connection exists between one's behaviour and the outcomes obtained, passivity, poor self-esteem, and depression may develop.

- **Efficacy expectations** that pertain to personal competence to achieve certain work outcomes. For instance, a teacher might burn out because he/she feels that he/she lacks the personal competence necessary to teach adequately. This differs from the outcome expectation, mentioned above, in that no matter what effort is made the result cannot be influenced. Lack of personal accomplishment, a hallmark of burnout, results from wrong efficacy expectations.

In addition to these three types of expectations, Meier (1983) introduced the notion of **contextual processing** in order to emphasise that these three types of expectations are influenced by individual and social factors. For instance, particular cognitive skills such as memory, but also personal beliefs and social norms, shape the individual's expectations.

Meier's (1983) cognitive-behavioural approach, grounded in learning theory, has not been studied empirically. However, it illustrates on a theoretical level that different kinds of expectations may play a different role in the development of burnout. More specifically, his approach predicts that wrong outcome expectations may trigger learned helplessness, a condition akin to burnout, whereas wrong efficacy expectations may affect the individual's sense of accomplishment.

Burnout as a disturbed action pattern

Building on German action theory, Burisch (1989, 1993) proposed an approach to burnout in which disturbed action patterns played a key role. Action theory considers so-called action episodes as the basic units of analysis. These episodes may range from minutes (a phone call) to decades (establishing a career) and they are hierarchically nested (the phone call may be instrumental for one's career advancement).

An undisturbed action episode is a cyclical process that starts when an individual's latent motives are activated: for example, when a psychologist has the opportunity to conduct a burnout workshop (see Figure 5.1).

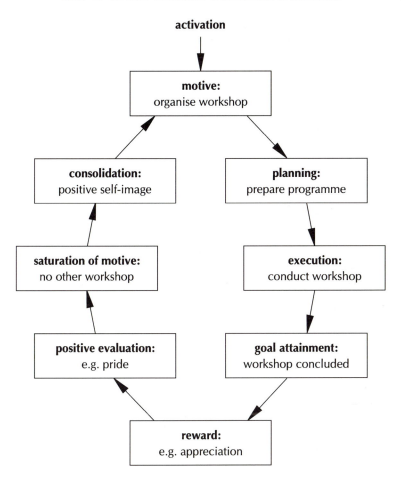

Figure 5.1 An undisturbed action episode (after Burisch, 1989, p. 78)

The workshop has to be carefully planned: a programme must be drafted, funds must be raised, announcements made, and so on. Next, the workshop is conducted and the goal is attained when the workshop is concluded. Rewards may come in terms of praise, appreciation, success rate, money, and so on. If it all works out well our psychologist will evaluate his or her efforts positively and experience corresponding emotions such as satisfaction and pride. For the time being, the original motive is satiated and there is no immediate need to conduct another workshop. The psychologist's self-image as a successful professional is consolidated. This successful action episode makes it more likely that in future our psychologist will be motivated to organise similar workshops. So much for a smooth undisturbed action episode.

However, many things can go wrong. Burisch (1989, 1993) distinguishes four patterns of disturbances that may disrupt an action episode:

■ **Motive thwarting** An obstacle may block goal attainment. In our previous example, the workshop may be so poorly designed that most participants drop out. Other examples are: the nurse whose patients are dying and the psychotherapist whose clients do not finish their treatment.

■ **Goal impediment** The goal can only be attained at the cost of unexpectedly high additional efforts. For instance, many unforeseen bureaucratic hurdles have to be taken before a burnout workshop can be conducted.

■ **Insufficient reward** The goal may be attained but the reward fails to live up to expectations. For instance, workshop participants do not show appreciation. In the burnout literature numerous examples of insufficient rewards can be found that are not proportional to the professional's efforts – unsatisfied or aggressive patients, unmotivated pupils, demanding customers.

■ **Unexpected negative side effects** Such effects may offset much or all that is gained. For instance, some participants may break down during the workshop and may have to be referred for psychiatric treatment.

Of course, a single disturbed action episode is not likely to cause burnout, rather it will lead to 'first-order stress': the usual tension that occurs when a particular situation or event taxes or exceeds the individual's adaptive resources. When first-order stress is not alleviated, and attempts to remedy the situation have repeatedly failed, the individual's sense of autonomy is threatened and 'second-order stress' develops. According to Burisch (1989), successful coping with second-order stress may lead to personal growth. Unsuccessful coping, on the other hand, may trigger the burnout process since action episodes will further deteriorate. For instance, action planning becomes more rigid or unduly sloppy, the individual's level of aspiration may shift downwards, the ability to accomplish goals is diminished, and the individual's self-image is profoundly shaken.

In a somewhat similar vein, Heifetz and Bersani (1983) conceived of burnout as a disruption of a cybernetic process. Instead of an action episode, they used a cybernetic model, which emphasises the role of feedback, as a basis for their analysis. Feedback in a cybernetic process makes it possible to monitor progress toward the goal and to change course whenever necessary. Heifetz and Bersani (1983) distinguish two major goals for human services professionals: client growth and professional growth. They argue that it is crucial for professionals to have objective and reliable milestones that accurately reflect the attainment of both goals. Furthermore, short-term indicators of progress must be available that reflect sequential steps en route to the long-term goals.

Unfortunately, these conditions are often not met in the human services. For instance, ambiguity exists about clear objectives of interventions (e.g. 'When is a patient cured?') and about standards of performance (e.g. 'What is adequate care?'). They state:

> Without 'operational definitions' of client growth and professional growth, the entire process of goal pursuit, feedback generation and course correction becomes open to uncertainty.   (Heifetz and Bersani, 1983, p. 54)

Moreover, short-term goals are often not specified and long-term goals are arbitrary or too distant. Most human service work involves facilitating the movement of clients along some developmental dimension: physical, cognitive, social or emotional. Therefore, 'final' goals involve arbitrary choices of points along these continua. By concentrating too heavily on these final (ambiguous and arbitrary) goals immediate goals are obscured or inflated. As a consequence, professionals may not register any significant progress and, wrongly, believe that they are incompetent.

Both approaches are useful in specifying various cognitive and motivational conditions in concrete action patterns that interfere with goal attainment at work. Unfortunately, they are not very specific as to why this may cause burnout and not, for example, dissatisfaction, or aggression.

Burnout as a loss of resources

Recently, a new general theory of stress termed Conservation Of Resources (COR) was applied to burnout (Hobfoll and Freedy, 1993; Hobfoll and Shirom, 1993). The basic tenet of COR theory is that people have a deeply rooted motivation to obtain, retain, and protect that which they value. Those things that people value are labelled resources and include: objects (e.g. tools, house); conditions (e.g. supportive social network, job stability); personal characteristics (e.g. social skills, self-esteem); energies (e.g. money, credit).

According to COR theory, psychological stress occurs when: resources are threatened (e.g. future job insecurity, role ambiguity); resources are lost (e.g. unemployment, divorce); resource gain does not follow appropriate investments of resources, so that a net loss of resources is experienced (e.g. failed promotion despite hard work). Generally, individuals deal with the resulting stress effectively by allocating or investing resources. For example, a new job is found through one's social network. This is, in fact, the reason why building up a pool of resources is essential: it acts as an insurance policy that allows the individual to better cope with future stressful events. After successful coping, a positive feedback loop is closed ('spiral of gain'): health increases and new resources are gained (e.g. new job) (see Figure 5.2).

However, coping may be unsuccessful and prolonged job stress or burnout might develop. COR theory defines burnout as '. . . a process of wearing out and wearing down of a person's energy, or the combination of physical fatigue, emotional exhaustion and cognitive wear-out that develops gradually over time' (Hobfoll and Shirom, 1993, p. 50). Burnout occurs when a net loss of valuable personal resources is perceived that cannot be replenished: physical vigour, emotional robustness, and cognitive agility. This net loss is experienced in response to job demands and cannot be compensated for by expanding other resources such as social support, self-esteem, or physical fitness.

In COR theory the symptoms as well as the causes of burnout are described in terms of resource loss. This illustrates the spiral of loss burned out individuals are caught in: because resources at work are threatened, lost, or not sufficiently gained (e.g. role ambiguity, interpersonal conflict, failed promotion) additional personal

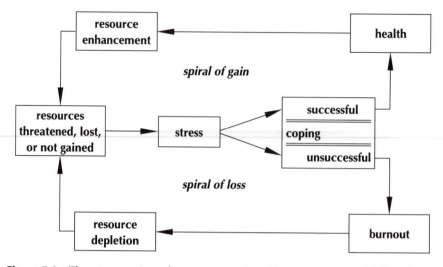

Figure 5.2   The conservation of resources model of burnout (after Hobfoll and Shirom, 1993, p. 56)

resources (e.g. stamina, social skills, network) have to be spent in order to cope successfully with the resulting stress. However, these resources are lacking with the result that coping must fail, burnout intensifies and resources are further depleted. Thus, essentially, the burned out individual is not able to allocate the appropriate resources because they are drained.

The COR approach to burnout emphasises that burnout is a self-perpetuating process and illustrates that the availability and use of resources are crucial for its development. The model as such has not been tested empirically but it was used as a theoretical framework for interpreting research findings from an intervention study (Freedy and Hobfoll, 1994).

Burnout as a narcissistic disorder

The psychoanalyst Fischer (1983) considered burnout to be a narcissistic personality disorder. He theoretically analysed three burnout cases whom he treated himself. According to Fischer (1983), individuals who idealise their jobs and suffer subsequent disillusionment might reduce their involvement by either reducing their ideals, or by leaving the situation. In both cases burnout would be avoided. However, neither option is acceptable to burnout candidates. Instead of giving up, reducing their ideals, or looking for other jobs, they redouble their efforts in order to achieve their unrealistic objectives.

The crucial question is: why do some people keep on working beyond reasonableness, common sense, and even concern for their own health and well-being? Fischer (1983) assumed that they ward off something that appears even more terrifying to them: losing their 'illusion of grandiosity'. The burnout candidate's basic sense of self-esteem is grounded in this narcissistic illusion: the erroneous notion of being special and superior. This illusion of grandiosity provides them with an intense and pervasive sense of pleasure, a 'high' analogous to drug addiction. Accordingly, literally everything is avoided that would threaten this illusion. For example, admitting that one's ideals have been unrealistic or that one has chosen the wrong profession. When a choice has to be made between giving up the illusion of grandiosity or exhausting one's resources, the burnout candidate unequivocally decides on the latter. Fischer (1983) disagreed with other authors such as Edelwich and Brodsky (1980), who considered the loss of ideals to be a hallmark of burnout. Instead he stated:

> It is not that these people feel that they have lost their ideal but, rather they are desperately trying to maintain it. It is not that they have been 'burned out', but that they are 'burning up'. (p. 44)

Recently, Glickauf-Hughes and Mehlman (1995) argued that in addition to Fischer's grandiose narcissist, its opposite, the depressive narcissist, is also likely to burn out. The latter is primarily characterised by an unstable self-esteem which essentially depends on emotional feedback from others rather than on the authenticity of his or her own feelings. They called this emotional dependence on others 'audience sensitivity' (p. 214) and argued that during childhood, narcissists develop an unusually powerful emotional antenna for detecting needs and feelings of others because they had to fulfil the audience role for their narcissistically disturbed parents. For these children, emotional responsiveness has been a matter of sheer psychological survival. This kind of sensibility that developed at the expense of an unstable self-esteem is precisely what qualifies them later as good and empathic helpers. According to the authors, it is not surprising that many helpers have been narcissistically abused as children. For helpers, an excellent emotional antenna is not only a strength, but it is also a weakness: being attuned to the needs and feelings of others is a demanding process, particularly for those whose self-esteem is unstable. Hence, a paradox exists: the good empathic helpers are in particular danger of burning out since their own needs are not likely to be satisfied by their often unresponsive and unmotivated client audience.

These psychodynamic interpretations of burnout, albeit speculative, point to the pivotal role of intrapersonal processes (retaining the illusion of one's grandiosity) and early childhood experiences (the development of audience sensitivity).

Burnout as an imbalance between conscious and unconscious functions

An alternative psychodynamic theory of the Swiss psychologist Carl Gustav Jung (1875–1961) has been applied to burnout by Garden (1991). According to Jung, two opposite personality types can be distinguished: 'feeling types', who are tender-minded and are characterised by concern and awareness for people; 'thinking types', who are hard-boiled, achievement-oriented and tend to neglect others. In fact, these two types represent psychic functions that are simultaneously present in each individual. One of these functions is usually preferred and the ego has become identified with that preferred, and thus conscious, function. The opposite, less preferred function, remains largely unconscious.

People tend to choose jobs that are compatible with their personality type. Research has shown that in health related, counselling, and educational fields the proportion of feeling types to thinking types is 4:1 (Garden, 1991). Obviously, feeling types prefer to work in the human services. In contrast, in occupations such as engineering, or in management predominantly thinking types are encountered, again in a similar proportion of 4:1. Garden (1991) argued that because of this confounding of occupation and personality, results from burnout research, predominantly conducted in the human services, are heavily biased towards feeling types.

Garden (1991) claims that her empirical studies (Garden, 1985; 1987; 1988; 1989) illustrate three psychodynamic principles:

- **A process of reversal** Burnout (energy depletion) is accompanied by a decrease in the characteristic that is normally associated with that particular type. That is, in burned-out feeling types, concern for others is diminished, whereas in burned-out thinking types lower levels of ambition and will to achieve are observed.

- **A process of convergence** Opposite types differ when they are low on burnout and are similar when they are high on burnout. That is, concern for others (will to achieve) is higher for feeling (thinking) types only at low levels of burnout but not at high levels.

- **Burnout occurs when the person fits to the job** Theoretically, feeling types are expected to be better at handling emotional demands, whereas thinking types are assumed to cope better with mental demands. Yet, it was observed that regardless of the type of job, emotional demands predict burnout in feeling types, whereas mental demands are predictive of burnout in thinking types. Thus, burnout is most strongly associated with the kind of demand each type is 'naturally' adapted to.

How should these findings be explained? In Jungian terms, if the psyche is deprived of energy, for whatever reason, the predominant conscious function drops into the unconscious and the opposite repressed function emerges. A feeling function sinking back into the unconscious would prevent a feeling type of person from relating to others in the usual concerned and pleasant way. At the same time, the repressed unconscious thinking function may emerge in a disruptive, negative way (e.g. depersonalisation). Similarly, in the thinking type, the achievement orientation crumbles and is partly replaced by tender-mindedness. This phenomenon of 'dropping into the unconscious' would explain the processes of reversal and convergence since both processes suggest that what was previously a conscious function is no longer conscious.

The observation that burnout is related to the fit between job and personality type and not to lack of fit, is explained by a similar psychodynamic self-regulatory process. According to Jungian theory, relying too much on one function creates an imbalance in the psyche that is counteracted by a similar increase of its opposite in the unconscious sphere. Thus, relying too much on one's feeling functions has the paradoxical effect of fuelling the unconscious thinking reservoir (and vice versa). So, when the repressed function emerges, its negative effect will be all the more devastating. Finally, according to Jung, withdrawal of functions from the conscious sphere and emergence of unconscious functions deprives the psyche of energy:

> In other words, the loss of use of one's conscious functions would create exactly the same experience as that used to describe burnout.   (Garden, 1991, p. 85)

To sum up: the dropping of conscious functions into the unconscious and the simultaneous emergence of actively repressed unconscious functions are both likely to occur when the psyche is deprived of energy – that is, when the individual is under stress. These are energy consuming processes that further deplete the individual's resources so that a vicious cycle is likely to develop. Furthermore, over-reliance on one psychic function increases the likelihood of the negative impact of its unconscious counterpart. Thus, in addition to the regulation of the imbalance between conscious and unconscious psychic functions, the Jungian perspective on burnout emphasises that burnout develops differently in different kinds of personality types.

Burnout as a failed quest for existential meaning

Drawing upon existential psychology, Pines developed a motivational approach to burnout in which the individual's basic need for meaning and significance plays a crucial role (Pines and Aronson, 1988; Pines 1993). The model, as depicted in Figure 5.3, has not been tested. Instead, Pines used it to interpret her research findings and to integrate her observations from the hundreds of burnout workshops that she had conducted (see Box 6.8).

The underlying assumption of the model is that only highly motivated individuals burn out:

> In order to burn out, one has first to be 'on fire'. A person with no such initial motivation can experience stress, alienation, depression, an existential crisis, or fatigue, but not burnout.   (Pines, 1993, p. 41)

Essentially, according to Pines, burnout is the final result of a gradual process of disillusionment in the quest to derive a sense of existential significance from work.

> Idealistic people work hard because they expect their work to make their lives matter in the larger scheme of things and give meaning to their existence.   (Pines, 1996, p. 83)

So, in essence, employees burn out because their experiences do not match their intentions and expectations.

More specifically, Pines (1993) proposed a fairly straightforward model that consists of two feedback loops (see Figure 5.3). In the positive loop, expectations and experiences match so that success is achieved and existential significance is experienced. By contrast, in the negative loop expectations are frustrated, failure occurs, and burnout develops.

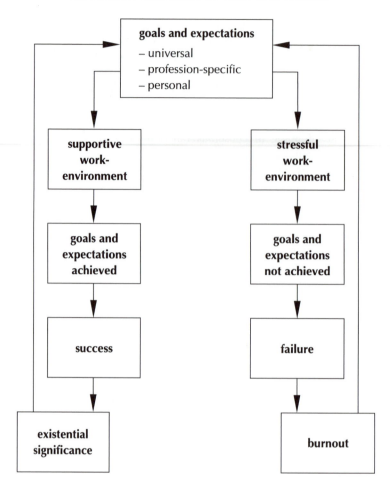

Figure 5.3   The existential model of burnout (*Source*: Pines, 1993, p. 42)

The existential approach distinguishes three types of goals and expectations that result from corresponding work motivations: universal motivations (e.g. to experience success and appreciation); profession-specific motivations (e.g. to have a significant impact on people, to help those in need); personal work motivations, that are based on an internalised romantic image of the job (e.g. the loving nurse from childhood). In combination, these three motivations constitute the expectation of significance, purpose and meaning of one's work.

Whether these expectations will be met depends on the work environment that can either be supportive or stressful. A supportive work environment provides a maximum of positive features that enables the professional to reach his or her goals by providing autonomy, support, and the necessary resources. It also minimises negative features that might interfere with goal-attainment, such as bureaucratic hassles, work overload, and role problems. By contrast, a stressful work environment is characterised by the lack of positive features and the presence of negative features. Indeed, a wealth of studies conducted by Pines (see Pines *et al.*, 1981; Pines and Aronson, 1988) as well as by others (see Chapter 4) documented the relationship between burnout and various work features, as predicted by the model.

It is important to note that in a case of burnout, by definition, a positive feedback loop existed initially: after all, at one time the professional had been 'on fire'. Gradually over time, for one reason or another, the original supportive work environment changed into a stressful and unsupportive environment. This might be caused by actual changes in the objective work situation (e.g. other duties or increased caseload) or by changed perceptions of that situation. In essence, the existential perspective on burnout illustrates that the professional's deeply rooted goals and expectations are instrumental for the development of burnout.

Conclusion

Virtually all individual approaches emphasise that a strong conscious or unconscious motivation (to help), including concomitant highly valued goals, expectations, and aspirations, is a necessary condition for the emergence of burnout. Furthermore, these approaches assume that these individual psychological characteristics often do not match the professional's experiences on the job. So a mismatch between intentions and reality exists. As a result of this poor fit, job stress occurs that may eventually lead to burnout when inadequate coping strategies are adopted and/or when the appropriate individual or organisational coping resources are lacking.

5.2 INTERPERSONAL APPROACHES

From the outset, emotional strains resulting from daily interactions with demanding, difficult or troubled recipients were considered to be the root cause of burnout, although this assertion is not fully supported by empirical research (see Chapter 4). The five interpersonal approaches that are discussed in this section try to unravel interactions at the workplace and to understand their relevance for the development of burnout. In a sense, the first approach marks the transition between individual and interpersonal approaches since the professional's competence, an individual characteristic, is considered in the social context of the helping relationship.

Burnout as a lack of social competence

The social competence approach assumes that burnout is inversely related to perceived competence and effectiveness in interpersonal relations with recipients (Harrison, 1983). The model, as depicted in Figure 5.4, bears close resemblance to the existential model because it also includes similar feedback loops: a positive loop leading to higher motivation and a negative loop leading to burnout. However,

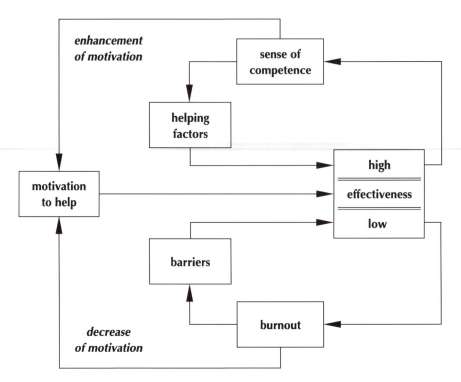

Figure 5.4   The social competence model of burnout (*Source*: Harrison, 1983, p. 31)

unlike the existential approach, social competencies rather than goals and expectations are the focal point:

> Social competence refers to how one feels about one's capacity to interact with, and therefore to influence the social environment.   (Harrison, 1983, p. 35)

The professional's motivation to help lies at the core of the model:

> The possibility of achieving competence as a helper – that is, knowing that one is having beneficial effects in a selected part of the social environment – seems to be the predominant motivation for helpers.   (Harrison, 1983, p. 32)

When the sense of having valued effects is lost, burnout may result. Like most previous authors, Harrison (1983) acknowledges that a misfit between intentions and reality reduces the likelihood of achieving the desired effects and thus fosters burnout. However, a proper motivation to help is not enough to be successful and effective in achieving one's goals. In addition, certain environmental conditions (e.g. clear and attainable goals) and individual characteristics (e.g. skills) are needed. These 'helping factors' boost performance and increase the professional's sense of social competence. By contrast, organisational obstacles or 'barriers' such as excessive paperwork, heavy caseloads, and lack of feedback might render one's efforts unsuccessful and may lead to burnout.

Recently, Cherniss (1993) argued in a similar fashion that burnout is inversely related to professional self-efficacy: the belief that one is capable of exercising control over events that affect one's work as a professional. His conceptualisation of professional self-efficacy is quite differentiated and includes: the task domain – the mastery of technical aspects of the professional role; the interpersonal domain – the ability to work harmoniously with others at work; the organisational domain – the ability to influence social and political forces within the organisation. As in the social competence approach, Cherniss (1993) assumes that goal attainment increases professional self-efficacy and motivation, whereas failures may result in poor motivation and eventually may cause burnout. He particularly points to the importance of the organisational domain since broader social and political skills are often neglected.

In conclusion, burnout is likely to occur when few successes are achieved and many failures are experienced. Because of the lack of positive feedback, the development of social competence and professional self-efficacy is impeded so that when failures occur the risk of burning out increases. As such, the social competence model has not been empirically tested.

Burnout as an emotional overload

According to Maslach (1993), burnout includes three components: emotional exhaustion is considered to be the stress component, whereas depersonalisation tries to capture a dimension of interpersonal relations, and reduced personal accomplishment reflects a dimension of self-evaluation. Maslach (1993) considered burnout to be a negative individual experience that is embedded in the context of interpersonal relationships at work and that involves the professional's conception of both self and others. She wrote about her initial approach that '. . . my social psychological background led me to frame the issue of burnout in terms of the social relationship between two people: one who gives, and the other who receives' (p. 29). Not surprisingly, Maslach studied burnout exclusively among helping professionals such as social welfare workers, physicians, poverty lawyers, police officers, teachers, and ministers. Recently, she proposed a more general approach to burnout that is not restricted to the human services and in which the organisational environment plays a more prominent role (Maslach and Leiter, 1997). This approach is discussed in the next section.

Figure 5.5 schematically represents Maslach's initial view on the development of burnout that can be deduced from her early writings (Maslach, 1982a; b).

Interpersonal demands resulting from the helping relationship are considered to be the root cause of burnout:

> Emotional exhaustion refers to feelings of being emotionally overextended and drained by one's contacts with other people.   (Maslach and Jackson, 1984a, p. 134)

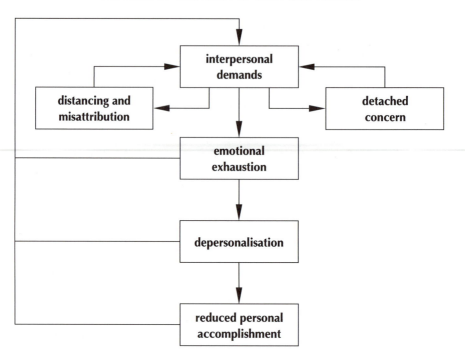

Figure 5.5   Emotional overload and burnout (see Maslach, 1982a; b)

What makes these contacts so demanding? Maslach (1982a; b) argued that these contacts are inherently difficult and upsetting because human services professionals deal with troubled people who are in need. In other words, helping relationships are emotionally charged by their very nature and therefore they inherently constitute a heavy psychological burden. More specifically, Maslach (1982b) identifies heavy caseloads and long and continuous direct contact with recipients as the main quantitative demands, whereas lack of support and poor skills to handle repeated, intense, emotional interaction with people are considered the principal qualitative demands. In addition to these crucial interpersonal stressors, other job related factors might also play a role, for instance, scarcity of resources, administrative red tape, or lack of feedback.

In order to deal with these emotional demands and perform efficiently and well, professionals may adopt techniques of detachment. Maslach (1982b) argued that by treating one's recipients in a more remote, objective way, it becomes easier to do one's job without suffering strong psychological discomfort. A functional way to do this is to develop an attitude of detached concern: the medical profession's ideal blending of compassion with emotional distance. This most preferred professional attitude is quite difficult to achieve because of its almost paradoxical nature: distancing oneself from people while helping them. In other words, a precious balance has to be kept between genuine concern and detached objectivity. Unfortunately,

many human services professionals cross the thin line between professional objectivity and distance, and unprofessional and dysfunctional detachment.

Maslach (1982b) identifies several distancing techniques that are used to reduce
interpersonal stress in interactions with recipients: using derogatory labels ('they're
just animals') and professional jargon ('a case of parentification'), intellectualisa-
tion ('he's exhibiting a delusional syndrome'), sick humour, physical withdrawal
(standing further away, avoiding eye contact), and psychological withdrawal (hiding
behind rules, avoiding tasks). In fact these techniques may be considered 'dehumanisation in self-defence' (Zimbardo, 1970). This notion refers to a process of
protecting oneself from overwhelming emotional feelings by responding to other
people more as objects than as persons. In addition to distancing, Maslach (1982b)
distinguishes a second dysfunctional way of dealing with interpersonal demands:
explaining situational stress in dispositional terms. In other words, when things go
wrong professionals may blame people rather than their work environment: either
the recipient is blamed ('blame the victim'), or the professional blames her- or
himself ('mea-culpa reaction'). Needless to say, using distancing techniques and
making false inferences or misattributions are likely to deteriorate the professional's
relationship with his or her recipient. Accordingly, instead of reducing emotional
demands, interpersonal stress increases and emotional resources are further depleted.

A hallmark of the burnout syndrome is a shift in the professional's perception of
recipients from a positive humanised pole to a negative and dehumanising one. This
is exactly what occurs when occasional distancing and/or misattribution become a
habitual pattern. In that case depersonalisation develops: a persistent callous, indifferent and cynical perception of recipients. According to Maslach (1982b), the
structure of the helping relationship is such that it promotes and maintains a negative perception of recipients and fosters depersonalisation. For instance, in the
human services the focus is on the recipients' problems rather than on their positive
aspects. Moreover, there is a lack of positive feedback: only when things go wrong
do professionals receive feedback from recipients and not when things work out
right. Finally, recipients are not always responsive to professionals: they fail to
follow their advice or guidance, or they drop out. These examples show that depersonalisation is likely to be reinforced in human services work.

As a result of these persistent negative perceptions of recipients, quality of care
or service is impaired. Depersonalised professionals are unable to perform adequately
because their major vehicle for success – compassion with and concern for others
– has been destroyed in an attempt to protect their own psychological integrity.
Because success is more and more lacking the professional's sense of personal
accomplishment erodes and feelings of insufficiency and self-doubt develop. As we
have seen in the previous approach such a decline of professional competence is a
self-perpetuating process.

In sum, Maslach (1982a; b) assumes that burnout is a sequential process that
starts with emotional exhaustion resulting from the emotional demands of dealing
with recipients. Next, in an inappropriate attempt to cope with exhaustion, depersonalisation develops. Because this is a dysfunctional coping strategy that further

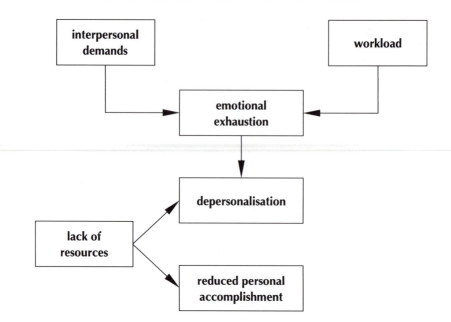

Figure 5.6   Leiter's process model of burnout (after Leiter, 1993, p. 245)

deteriorates the relationship with recipients, more and more failures are experienced so that gradually a sense of diminished personal accomplishment develops.

What about the empirical support for this model? To date, the full model has not been tested empirically, rather it has served as a heuristic for investigating the sequence of the three burnout dimensions and the relationships between various stressors and these dimensions. Unfortunately, the effects of distancing techniques, misattributions and detached concern have not really been examined. Recent evidence suggests that the three factor sequential model as conceived by Maslach is slightly superior to models that assume an alternative sequence such as the model of Golembiewski and his colleagues (see later) (Lee and Ashforth, 1993; Cordes *et al.*, 1997). Based on the approach of Maslach, Leiter (1990; 1991; 1993) conducted a series of studies in which he distinguished quantitative job demands (e.g. work overload, hassles), qualitative job demands (e.g. interpersonal conflict), and lack of resources (e.g. lack of social support, lacking opportunities for skill enhancement, poor client cooperation, lack of autonomy, and participative decision making). Demands were expected to be related to emotional exhaustion, whereas resources were expected to be related to depersonalisation and lack of personal accomplishment. Indeed, as shown in Figure 5.6, the results largely confirmed his hypothesised model, albeit the data were based on cross-sectional research.

Unlike the initial model (Figure 5.5), reduced personal accomplishment seems to develop rather independently from both other burnout dimensions. Recently, Lee and Ashforth (1996) presented a meta-analysis that includes over 60 studies and found that emotional exhaustion is particularly related to job demands (e.g. role problems, work load, work pressure, stressful events, unmet expectations), whereas poor personal accomplishment is related to lack of resources (e.g. poor social support). Depersonalisation appears to be related to both job demands as well as lack of resources (see also Chapter 4). Essentially, these findings are in agreement with Figure 5.6 which shows that indirectly depersonalisation is a function of poor job demands. Lee and Ashforth (1996, p. 128) conclude that:

> The results [of the meta-analysis: *the authors*] are also consistent with Leiter's (1993) belief that personal accomplishment develops largely independently of emotional exhaustion and depersonalisation.

In Chapter 4, however, we showed that job-related stressors such as workload are much stronger and more consistently associated with emotional exhaustion than interpersonal, client-related stressors.

Recently, an indication was found of the presence of a negative feedback loop as assumed in the model of Maslach (see Figure 5.5). A longitudinal study among general practitioners confirmed earlier cross-sectional findings (Van Dierendonck et al., 1994) indicating that the attitudinal component of burnout (a combination of depersonalisation and reduced personal accomplishment) negatively affects the quality of the doctors' relationship with their patients (Bakker et al., 1997). In turn, poor interpersonal patient relationships predicted the doctors' level of emotional exhaustion and the concomitant negative attitudes across a period of 5 years.

In sum: empirical findings support the notion that burnout is a multidimensional construct whose dimensions are differentially related to job demands and resources at work. Emotional exhaustion develops in reaction to job demands, including interpersonal demands, and seems to lead to depersonalisation. Personal accomplishment is positively influenced by the presence of resources and largely develops in parallel to both other dimensions.

Burnout as a lack of reciprocity

Recently, Buunk and Schaufeli (1993) made an attempt to link burnout with social exchange processes at the interpersonal level. Their central thesis is that burnout develops primarily in the social and interpersonal context of the work organisation and that, in order to understand its development, attention has to be paid to the way individuals perceive, interpret, and construct the behaviours of others at work. They distinguish two aspects of social exchange processes that are relevant for burnout: reciprocity and symptom contagion. In recent years, both principles have been further developed as separate approaches to burnout.

Buunk and Schaufeli (1993) follow Maslach's (1993) notion of burnout as a multidimensional syndrome that is rooted in the emotionally demanding interpersonal relationship between caregiver and recipient. By definition, this relationship is complementary in the human services. This is semantically well-illustrated by the terms 'caregiver' and 'recipient'. The former is supposed to *give* care, assistance, advice, support, and so on, whereas the latter is supposed to *receive* it. Nevertheless, professionals look for some rewards in return for their efforts. For example, they expect the recipients of their care to show gratitude, to improve, or at least make a real effort to get well. However, these expectations are seldom fulfilled because, for instance, recipients do not improve because they suffer chronically and they take the professional's efforts for granted – they are paid for it, after all. Hence, it is likely that over time a lack of reciprocity develops, whereby caregivers feel that they continuously put much more into relationships with their recipients than they receive back. As Buunk and Schaufeli (1998) have pointed out, reciprocity plays a central role in human life and establishing reciprocal social relationships is essential for the individual's health and well-being. They argue that the strong universal preference for reciprocal interpersonal relationships is deeply rooted since it may have fostered survival and reproductive success in our evolutionary past. Lack of reciprocity, in their evolutionary view, not only leads to negative emotions, but it also motivates to restore reciprocity.

Although the importance of reciprocity has been recognised throughout the burnout literature, it has not been systematically included as a theoretical approach. For example, Farber (1983, p. 6) concluded in the introduction of one of the first and most influential edited volumes on burnout that: 'Burnout, then, can be viewed as a process that occurs when workers perceive a discrepancy between their input and expected output. To balance the equation, burned out workers begin to give considerably less to their jobs'. In a similar vein, Heifetz and Bersani (1983) stated: 'It is not the heavy emotional investment per se that drains the provider, rather it is an investment that has insufficient dividends'. Cherniss (1995) who conducted a unique 12 year follow-up study among young helping professionals starting when they entered their careers noted:

> Most of the new professionals lost their idealism very early in their careers, and this change seemed to be caused by the imbalance between what the professionals put into their jobs and what they got out of them.   (p. 46)

In their argument, Buunk and Schaufeli (1993) drew heavily upon equity theory, probably the most influential social exchange theory that has been applied to interpersonal relationships as well as to the relationship of the employee with the organisation. According to equity theory, people pursue reciprocity in interpersonal and in organisational relationships: what they invest and gain from a relationship should be proportional to the investments and gains of the other party in the relationship. When individuals perceive relationships as inequitable they feel distressed and they are strongly motivated to restore equity. Or as Freudenberger and Richelson (1980, p. 175) have put it: 'Since burnout sets in when the effort spent is in inverse proportion to the reward received, it becomes imperative to balance the equation'.

More specifically, lack of reciprocity – an unbalanced helping relationship – drains the professionals' emotional resources and eventually leads to emotional exhaustion. Initially, when the expected outcomes do not occur, care-givers are likely to invest more in their relationships with recipients. That is, they spend more time and pay extra attention to their recipients. When this does not pay off in terms of better outcomes, the imbalance increases and resources are further depleted.

Investing in a relationship without receiving appropriate outcomes is highly energy consuming and extremely frustrating for most people. The resulting emotional exhaustion is typically dealt with by reducing the investment in the relationships with recipients. That is, by responding to recipients in a depersonalised way instead of expressing genuine empathic concern. Hence, according to Buunk and Schaufeli (1993) depersonalisation can be regarded as a way of restoring reciprocity by withdrawing psychologically from recipients. However, this way of coping with an unbalanced interpersonal relationship is dysfunctional since it deteriorates the helping relationship, increases failures and thus fosters a sense of diminished personal accomplishment.

As expected, positive relationships were found between lack of reciprocity at the interpersonal level and all three dimensions of burnout in several occupational groups such as student-nurses (Schaufeli *et al.*, 1996b), general hospital nurses (Schaufeli and Janczur, 1994), critical care nurses (Schaufeli and Le Blanc, 1997), general practitioners (Van Dierendonck *et al.*, 1994; Bakker *et al.,* 1997), and correctional officers (Schaufeli *et al.*, 1994b). Furthermore, it appeared that the relationship between lack of reciprocity and burnout is moderated by individual differences. Among nurses who are high in communal orientation – those who have a strong desire to give in response to the needs and out of concern for others – an unbalanced relationship hardly matters, whereas for those low in communal orientation it was clearly related to burnout (Van Yperen *et al.*, 1992; Van Yperen, 1996). Apparently, nurses who have a strong desire for reciprocity in their relationships with patients are in particular danger of burning out. In other words, those who expect too much tend to burn out, a result that is consistent with most individual approaches discussed in the previous section.

As noted before, equity theory can also be applied at the organisational level as well. Similar social exchange processes to those that are observed in interpersonal relationships govern the relationship of the professional with his or her organisation. Therefore, Schaufeli *et al.* (1996b) have proposed a dual-level social exchange model (see Figure 5.7).

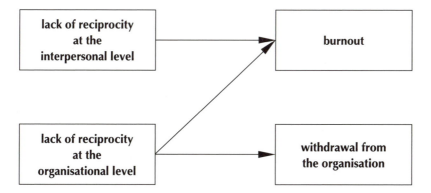

Figure 5.7   The dual-level social exchange model of burnout (*Source*: Schaufeli *et al.*, 1996b, p. 228)

They argue that in addition to an unbalanced relationship at the interpersonal level, burnout is also caused by a lack of reciprocity at the organisational level, that is, by a violation of the psychological contract. A psychological contract refers to the expectations held by employees about the nature of their exchange with the organisation (Rousseau and Parkes, 1993). Expectations concern concrete issues such as workload, as well as less tangible matters such as esteem and dignity at work, and support from supervisors and colleagues. Thus, the psychological contract reflects the employees' subjective notion of reciprocity: he or she expects gains or outcomes from the organisation that are proportional to his or her investments or inputs. When the psychological contract is violated because experience does not match expectations, reciprocity is corroded. Schaufeli *et al.* (1996b) argued that in addition to the usually observed cognitive and behavioural withdrawal reactions (dissatisfaction, reduced organisational commitment, turnover, and absenteeism), violation of the contract may also lead to burnout. This is in agreement with the literature that shows that poor organisational conditions such as pay inequity, lack of control, lack of role clarity, and lack of social support, are determinants of burnout (see Chapter 4). Such poor job conditions may in fact be considered to be aspects of a violated psychological contract because an employee expects adequate pay, job autonomy, a clear work role, and social support from the organisation.

From the perspective of equity theory, lack of reciprocity and the resulting distress that is experienced when the psychological contract is violated can be dealt with by decreasing one's investments. In that case, the organisation's outcomes are simultaneously lowered. For instance, depersonalisation not only indicates the professional's decreased readiness to invest emotionally in recipients, but it simultaneously lowers the organisation's outcomes since it is likely to deteriorate the quality of the provided services. So, essentially, the dual-level model considers burnout as the outcome of an inadequate strategy to restore equity at two levels by reducing one's involvement and commitment in recipients as well as in the organisation.

The dual-level model was tested successfully in two samples of student-nurses (Schaufeli *et al.*, 1996b). Recent studies among teachers (Van Horn *et al.*, in press), therapists from a forensic psychiatric clinic, and staff working with the mentally disabled (Van Dierendonck *et al.*, 1996) confirmed that burnout is related to perceptions of inequity at the organisational level. The validity of these findings is corroborated by similar results from a recent study that used a slightly different theoretical framework – the Effort-Reward Imbalance model (Siegrist, 1996). It appeared that, as expected, nurses who experienced an imbalance between the effort spent at their jobs and the rewards received in return had higher burnout levels than those who did not experience a similar imbalance (Bakker *et al.*, 1998a). Thus, once again, an unbalanced relationship with the organisation seems to be related to burnout.

In conclusion, it seems that a lack of reciprocity is a key concept for understanding burnout. Rather than simply working too long, too hard with too difficult recipients, as is assumed in most traditional models of burnout, it appears that the *balance* between investments and outcomes is crucial for the development of burnout. It appears that this mechanism works in similar ways at the interpersonal level of care-giver and recipients, and at the organisational level of employee and organisation.

Burnout as an emotional contagion

In early accounts of burnout, its contagious nature had already been described. For instance, Schwartz and Will (1953, p. 348), who presented the case study of Miss Jones (see Box 1.2), wrote that: 'Miss Jones gradually became more discouraged, so that by the time of the first week she was sharing the feelings and attitudes of the other staff-members and functioning in the same ineffective way'. Edelwich and Brodsky (1980, p. 25) used an analogy with staphylococci germs that spread the flu:

> Burnout in human services agencies is like staph infection in hospitals: it gets around. It spreads from clients to staff, from one staff member to another, and from staff back to clients. Perhaps it ought to be called 'staff infection'.

Unlike the previous approaches, the principle of emotional contagion cannot be used to explain why burnout develops, rather it helps to explain the mechanism by which it spreads once it has developed.

The first empirical indication for the contagious nature of burnout comes from Rountree (1984) who studied over 180 task groups in over 20 different work settings. He found that almost 90% of those high in burnout were members of work groups having at least 50% of all their members suffering from advanced burnout. Golembiewski *et al.* (1986) concluded after reviewing similar additional studies that: 'Very high and very low scores on burnout tend to concentrate to a substantial degree' (p. 184). They add: 'These findings suggest "contagion" or "resonance" effects' (p. 185). Of course, this concentration of burnout in particular groups may also be explained by higher workloads in these groups, which would contradict a symptom contagion explanation. However, this alternative hypothesis was recently rejected in a study that included almost 80 European intensive care units (Bakker *et al.*, 1998b). It appeared that, after controlling for job autonomy, subjective workload and objectively assessed workload (i.e. complexity of nursing tasks), nurses' levels of experienced burnout as well as levels of observed burnout among their colleagues of the unit differed systematically across units. Thus, independently from the nurses' workload, levels of burnout were higher in some units compared with other units, and nurses from these units observed more burnout complaints among their colleagues than their fellows did in the other units. This intriguing result supports the contagion hypothesis of burnout.

How can this contagion of burnout be explained? Buunk and Schaufeli (1993) have suggested that colleagues may act as models whose symptoms are imitated through a process of emotional contagion. That is, individuals under stress may perceive symptoms of burnout in their colleagues and automatically take on these symptoms. Emotional contagion is defined as: 'The tendency to automatically mimic and synchronise facial expressions, vocalisations, and movements with those of another person, and consequently, to converge emotionally' (Hatfield *et al.*, 1994, p. 5). The emphasis in this definition, supported by myriad laboratory and field studies, is clearly on non-conscious emotional contagion. There is, however, an alternative way in which people may catch emotions from others. Contagion may also occur through a conscious cognitive process by 'tuning in' to the emotions of others. This will be the case when an individual tries to imagine how (s)he would feel in the position of another, and, as a consequence, experiences the same feelings. The professional attitude of human services workers that is characterised by empathic concern is likely to foster such a process of consciously tuning in to the emotions of others. It is important to note that negative moods appear to be more contagious than positive moods, suggesting that, similar to depression, burnout symptoms would be likely candidates for emotional contagion.

Few studies suggest that emotional contagion may play a role in the development of burnout. Westman and Etzion (1995) studied about 100 male military officers and their wives and found that burnout transferred from husbands to wives and vice versa. In another study among professionals who work with the homeless it was found that emotional contagion was directly as well as indirectly, through communicative responsiveness, related to burnout (Miller *et al.*, 1995). Recently, Bakker *et al.* (1997) observed that general practitioners who perceived burnout complaints among their colleagues reported higher levels of emotional exhaustion and subsequent negative attitudes (depersonalisation and reduced personal accomplishment) than those who did not perceive such complaints. In addition, the doctors' individual susceptibility to emotional contagion was positively related to burnout, particularly in combination with the perception of burnout symptoms in their colleagues. That is, practitioners who perceived burnout complaints among colleagues and who were susceptible to burnout reported the highest exhaustion scores. A somewhat similar result was found in an earlier study among nurses by Groenestijn *et al.* (1992) who included instead of susceptibility to contagion another individual difference variable: the need for social comparison. They found that nurses who perceived that many of their colleagues showed burnout symptoms expressed higher levels of emotional exhaustion, especially when they had a strong need to learn more about others in a similar situation. Thus, social comparison with similar others is associated with higher burnout levels (see also Buunk *et al.*, 1994).

In conclusion, the observation that burnout seems to spread like an infectious disease might be explained by assuming a process of emotional contagion through which professionals take on burnout symptoms (particularly emotional exhaustion) that are observed among others who are in a similar position. This perspective emphasises that not only interpersonal relationships with recipients, but also interpersonal relationships with colleagues may contribute to burnout.

Burnout as an emotional labour

In contrast to the previous interpersonal approaches, the perspective of emotional labour has not explicitly been applied to burnout. Yet, we feel that the notion of emotional labour is potentially important for our understanding of burnout since it captures an essential aspect of the professional-recipient relationship. Emotional labour is defined as the act of displaying the appropriate emotion. Or more precisely as: '. . . the effort, planning and control needed to express organisationally desired emotion during interpersonal transactions' (Morris and Feldman, 1996, p. 987).

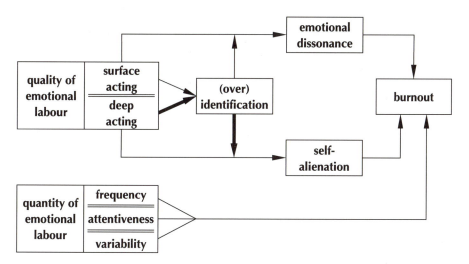

Figure 5.8 The emotional labour model of burnout

Thus, nurses are expected to display compassion, psychotherapists understanding, teachers enthusiasm, correctional officers rigour, and funeral directors sadness. In other words, occupation- or organisation-specific standards exist that dictate when and how emotion should be expressed. These standards are referred to as feeling rules or display rules.

The sociologist Arlie Hochschild (1983) was the first to use the concept of emotional labour to analyse the jobs of service professions. In her book *The Managed Heart: Commercialisation of Human Feeling* she presents a vivid and in-depth picture of emotional labour performed by flight attendants and bill collectors. Hochschild (1983) distinguished between two types of emotional labour:

- **Surface acting** involves simulating emotions that are not actually felt, through verbal and non-verbal cues such as facial expressions, gestures, and voice tone. Thus, in surface acting, emotions are feigned that are not experienced.

- **Deep acting** involves attempts to actually feel the emotions that one wishes to display. Basically, two avenues may be used for deep acting: exhorting feelings (i.e. actively evoking or suppressing emotions) or trained imagination (i.e. actively evoking thoughts, images and memories to induce the associated emotion – e.g. thinking about holidays to feel relaxed).

Both types of acting may have severe negative consequences (see Figure 5.8).

Surface acting, portraying emotions that are not felt, creates a sense of strain that is called emotional dissonance. The conflict between genuinely felt emotions and emotions required to be displayed may cause burnout because it depletes the individual's emotional resources (emotional exhaustion) and it fosters cynicism, withdrawal, and detachment (depersonalisation). According to Ashforth and Humphry (1993), the latter has to be interpreted as a maladaptive defence mechanism that is utilised to ameliorate the emotional strain resulting from dissonance. Deep acting, on the other hand, may lead to self-alienation as one is in danger of losing touch with one's authentic self – role playing becomes role taking. Eventually, it may impair the ability to express one's genuine emotions. Since deep-acting requires excessive energy, it may in the long run deplete one's emotional resources. In addition, similar maladaptive defence mechanisms may be used in order to reduce the strain resulting from self-alienation.

Recently, Morris and Feldman (1996) have extended the emotional labour approach by arguing that in addition to qualitative aspects of emotional labour (emotional dissonance and self-alienation) its quantitative aspects have to be considered:

- **frequency** of emotional display (i.e. the frequency of interactions between professionals and recipients);
- **attentiveness** to required display rules (i.e. the duration and intensity of emotional labour – a brief and superficial greeting of a flight attendant versus the long and intense care of an oncology nurse);
- **variability** of expressed emotions (i.e. the range of emotions that a professional is expected to display – the cheerful and friendly flight attendant who should only express positive emotions versus the teacher who in addition to positive emotions should also express negative emotions to support discipline and neutrality to demonstrate fairness).

According to Morris and Feldman (1996), when frequency, attentiveness and variability are high, more emotional labour is required and the risk of burning out increases. This is consistent with Maslach (1982a; b) who identified the exposure to emotional demands in the helping relationship as the root cause of burnout.

Ashforth and Humphrey (1993) argued that, basically, emotional labour is a functional strategy to adapt to one's role in an organisation by complying with the prevailing display rules. That is, emotional labour may increase self-efficacy and task effectiveness because it regulates and smooths interactions with recipients. Furthermore, by displaying the appropriate emotion instead of expressing the authentic emotion, the professional creates a certain distance that allows objectivity and is able to maintain his or her emotional equilibrium. In fact, this is what happens in detached concern.

Obviously, emotional labour may have positive as well as negative effects. Ashforth and Humphrey (1993) argue that the extent to which the professional identifies with his or her work role is crucial for the effect of emotional labour. Compliance with display rules through surface acting, but particularly through deep acting, will over time foster identification with the work role. The more strongly one identifies with the work role, the more positive the impact on well-being since emotional dissonance and self-alienation are reduced. However, identification carries an emotional risk:

> It may bind one to the role such that one's well-being becomes more or less yoked to perceived successes and failures in the role. (p. 107)

This is exactly what happens in burnout: overidentification with one's work role so that too much emotional labour is required in order to achieve the highly stacked work goals. The risk of overidentification is particularly high in the case of deep acting.

In conclusion, the notion of emotional labour illustrates the dynamic nature of the interpersonal relationship with recipients. Burnout is likely to develop as a result of surface acting (when identification is low) and deep acting (when overidentification occurs). In addition, the sheer amount of emotional labour increases the risk of burnout. Unfortunately, to date no empirical evidence is available that convincingly documents the validity of the emotional labour approach for burnout.

Conclusion

Some interpersonal approaches describe burnout as a result of lacking social competence or as a sequential multifaceted reaction to emotional overload. Other approaches try to explain the development of burnout by pointing to underlying psychological processes such as social exchange (i.e. lack of reciprocity), emotional contagion, or emotional labour. In contrast to most individual approaches, interpersonal approaches are, at least partly, supported by empirical evidence.

5.3 ORGANISATIONAL APPROACHES

Three organisational approaches are discussed below that differ considerably in terms of content, scope, research method, empirical support, and general flavour. The first approach originates from a thoughtful, in-depth and largely qualitative analysis of a limited number of human services professionals who had just entered the field. All of them were re-interviewed 12 years later. By contrast to the first approach that is illustrated with vivid personal experiences of the involved professionals, the second approach is strictly quantitative. It builds on an impressive database of over 80 studies that include thousands of employees, some from outside the human services, in over 30 countries around the globe. The third approach is mostly descriptive; it tries to relate recent organisational changes in a number of areas to burnout.

Burnout as a 'reality shock'

At the end of the 1970s, Cherniss (1980a; b) interviewed 26 human services professionals (i.e. lawyers, high school teachers, public health nurses, and mental health professionals) who had just entered their fields. He observed massive disappointment among these novices. They failed to find what they originally sought for as professionals: their ideals, intentions and expectations had dramatically clashed with organisational reality. As a result of this reality shock the professionals in the study had gradually lost their sense of mission and zeal and showed a strong tendency to withdraw psychologically from their recipients and from their jobs. Cherniss (1980a) warns that a schematic model might lead to confusing simple

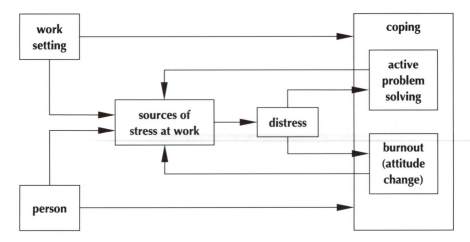

Figure 5.9   The reality shock model of burnout (Cherniss, 1980a, p. 208)

abstractions with complex reality he summarised his ideas in a straightforward model of early career development (Figure 5.9).

At the core of his model lie various sources of stress that result from the work setting and, to a lesser extent, from the person. These sources of stress and the resulting distress can be dealt with, either adequately by active problem solving, or inadequately by developing negative attitudes. The former reduces distress because its causes are removed or altered, whereas the latter increases distress so that ultimately burnout develops – the professional's energy is drained and negative attitudes have become a habitual pattern. Here we encounter the familiar positive and negative feedback loops. In the original model, the positive feedback loop was not recognised – i.e. the arrow from active problem solving to sources of stress. Recently, however, Cherniss (1993) argued that successful problem solving and achieving one's goals enhances professional self-efficacy, which, he argued, is the opposite of burnout.

More specifically, Cherniss (1980a, pp. 21–96) distinguishes five sources of stress that are to a large extent typical for novices in the human services: doubts about competence – despite many years of formal school training, novice professionals feel unprepared for their roles; difficult recipients – they lack motivation or ability, or are manipulative or even abusive; bureaucracy – rather than being the autonomous 'free agents', novices find themselves deeply involved in time consuming politics and paperwork; lack of stimulation or fulfilment – since novelty and challenge soon make way for oppressive routine and trivia, novices experience lack of variety, meaning, and intellectual discovery; lack of collegiality – rather than a source of help, support, guidance, and stimulation, peers turn out to be one more source of strain because of rivalry, conflict and differences in values. In fact, these five sources of stress at work result from the clash of the everyday reality of the job with the professional's initial expectations about competence, recipients, autonomy, self-realisation, and collegiality, respectively. In Chapter 1 we have seen that these five expectations constitute the so-called professional mystique (Cherniss, 1980a, pp. 249–256).

According to Cherniss (1980a, pp. 158–180), eight critical factors in human services work settings can be distinguished that produce stress and might cause burnout:

- **Poor orientation** Usually, a proper orientation process, sensitive and responsive to the novice's needs, is lacking. Many are immediately confronted with all of the demands imposed on the most experienced worker.

- **High work load** Only a short time can be spent on each recipient which reduces the chances that the professional's efforts will be successful. Besides, there is too little time to consult more experienced colleagues.

- **Routine** Much of the ordinary work of human services professionals involves routine matters.

- **Narrow scope of client contact** Usually, human services professionals are able to relate to only a few aspects of the recipient's situation (e.g. physicians focus on the patient's body, social workers on the family, and lawyers on legal matters).

- **Lack of autonomy** A set of rather strict rules and regulations governs the professional's behaviour and limits his or her autonomy on the job.

- **Incongruent institutional goals** The goals of the institution may not match personal values (e.g. a teacher favours an interdisciplinary approach that is not shared by the school).

- **Poor leadership and supervision practices** Novices, in particular, need support, guidance and proper feedback from supervisors.

- **Social isolation** In many human services, colleagues are neither physically nor psychologically available.

Cherniss discovered that the presence or absence of the eight previous factors in the work setting made the difference: when an orientation programme existed, workload was not too high, the job was stimulating, contacts with clients were broad, job autonomy existed, goals were congruent, leadership was adequate and social isolation was absent, professionals did not experience 'reality shock' and burnout. As Figure 5.9 indicates, aspects of the work setting also directly influence the way professionals cope with stress. For instance, if autonomy is lacking few, if any, active problem solving strategies can be used.

In addition to these organisational factors, two kinds of personal factors seem to be critical (Cherniss, 1980a, pp. 181–205). The first of these is the balance of supports and demands outside of work. For instance, stress and difficulty at home may impede adjustment at work. Also, the lack of a stable, close and available network of family and friends is likely to be associated with burnout. The second personal factor is personal differences in initial outlook matter. Based on his interviews Cherniss describes four so-called career orientations that new professionals brought to their jobs: *social activists* who want to do more than 'just' help their clients – they want to bring about social and institutional change; *artisans* who are intrinsically motivated and for whom professional growth and development are of prime importance; *careerists* who seek success as conventionally defined in terms of prestige, respectability and financial security; *self-investors* who are more involved in their personal lives outside work than in their careers – they typically work to live.

Not surprisingly, social activists and artisans run a greater risk of experiencing 'reality shock' and burning out than do careerists and self-investors. Their goals, intentions, and expectations are high and they are therefore more likely to fail. Indeed, Burke and Greenglass (1988) showed that teachers who described themselves as social activists experienced the most severe reality shock and had the highest burnout scores relative to the other career orientations. As Figure 5.9 indicates, person factors (i.e. situation outside work and career orientation) influence the professional's attempts to cope with distress. For instance, being preoccupied with family problems impairs active problem solving at work.

Finally, Cherniss (1980a, 97–133) described six attitude changes that are typical for burnout, in addition to the characteristic energy depletion: reduced aspirations, increased indifference, emotional detachment, loss of idealism, alienation from work, and increased self-interest.

Cross-sectional studies among police officers (Burke *et al.*, 1984) and teachers (Burke and Greenglass, 1989a) largely support the validity of the model. As expected, path-analysis showed two significant indirect paths from work setting characteristics and personal characteristics to burnout. Both were mediated by experienced sources of stress. In addition, significant direct paths were found from work setting and personal characteristics to burnout. Cross-sectional (Burke and Greenglass, 1989b) evidence among teachers corroborate these results by showing that sources of stress mediate the relationship between work setting characteristics and burnout. However, a longitudinal study (Burke and Greenglass, 1995) rendered contradictory results (see Chapter 4.4).

After 12 years Cherniss (1995) re-interviewed all 26 original respondents. Only eight of them (32%) still worked directly with recipients, whereas 13 (50%) had left the field. Roughly three groups could be distinguished. First, those who made a radical change (e.g. the poverty lawyer who became a Beverly Hills tax attorney). Second, those who managed to sustain the idealism that they had as novices. They had, in one way or another, successfully resisted the forces that undermined the commitment of their fellow novices. Third, those who recovered from burnout.

> They never completely regained the kind of idealism, caring, and commitment that they had when they began their careers. Once burned out, they were cautious about investing too much of themselves in either their careers or their work with clients.   (Cherniss, 1995, p. 12)

It appeared that those who were more burned out early in their careers were less likely to change careers and were more flexible in their approach to work (Cherniss, 1989). Further analysis showed that four common factors seemed important for their survival (Cherniss, 1990): a change to a more favourable work setting; the growth over time of professional self-efficacy; the development of special interests on the job; relatively greater vocational maturity at the beginning of their careers. Taken together, the in-depth follow-up study suggests that early career burnout does *not* lead to any significant negative long term consequences.

In conclusion, the model of early career burnout of Cherniss (1980a) carefully describes the dynamic process of imbalance, adjustment and change that occurs in human services professionals during the first stage of their careers. More particularly, the model specifies various aspects of work settings in the human services that might produce typical sources of stress, and eventually, may lead to the development of burnout.

Burnout as virulent process

The approach of Golembiewski and his colleagues is rather straightforward (Golembiewski and Munzenrider, 1988; Golembiewski *et al.*, 1996). They consider burnout as a 'virulent process' that develops progressively through eight phases. Burnout is triggered by job stressors and leads, in its turn, to physical symptoms, decreased performance and reduced productivity (see Figure 5.10). Accordingly, it has negative consequences not only for the individual but also for the organisation.

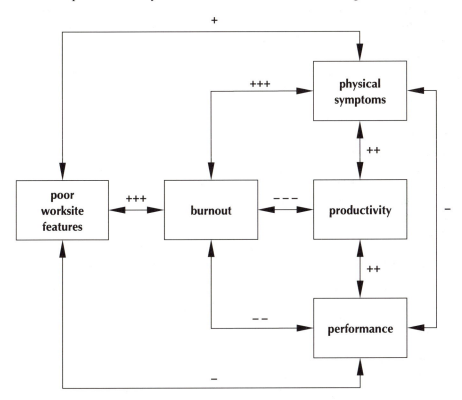

Figure 5.10 Strengths of relationships in the environmental model of burnout (after Golembiewski and Munzenrider, 1988, p. 109)
Notes: +++ or −−− = 'robust' association (i.e. > 15% shared variance); ++ or −− = 'moderate' association (i.e. 10–15% shared variance); + or − = 'modest' association (i.e. 5–10% shared variance)

Table 5.1   Progressive phases of burnout*

	I	II	III	IV	V	VI	VII	VIII
DEP	low	high	low	high	low	high	low	high
PA(r)	low	low	high	high	low	low	high	high
EE	low	low	low	low	high	high	high	high

Notes: **DEP** = MBI-depersonalisation; **PA(r)** = MBI-personal accomplishment (reduced); **EE** = MBI-emotional exhaustion.

* Golembiewski and Munzenrider (1988; p. 28): *Phases of burnout: Developments in concepts and applications.* Reprinted with permission of Greenwood Publishing Group Inc., Westport, CT.©

Although Golembiewski agrees with the three-dimensional nature of burnout as proposed by Maslach (1982a), he uses a rather more strongly modified version of the MBI. Most of the research of his team is conducted outside the human services in commercial organisations – '. . . burnout is where you find it: that is, everywhere . . .' (Golembiewski and Munzenrider, 1988, p. 12). Most importantly, Golembiewski disagrees with Maslach (1982a) about the sequence of the three burnout components. According to Golembiewski, the burnout process starts with depersonalisation, followed by a lack of personal accomplishment and emotional exhaustion, respectively. To date, Golembiewski and his colleagues have neglected to provide a convincing theoretical rationale for this perspective. Recent empirical studies, however, suggest that the sequential model of Maslach fits somewhat better to the data than the sequence proposed by Golembiewski, although the results are not entirely conclusive (Lee and Ashforth, 1993; Cordes *et al.*, 1997).

Dichotomising the distribution of the three MBI scale-scores at the median as 'high' and 'low' generates eight phases of progressive burnout (see Table 5.1). It is important to note that these phases do not represent a developmental process but a way of classifying individuals according to the 'virulence' of their burnout symptoms. In other words, those in the most advanced phases of burnout have not necessarily passed through the previous less virulent phases. In this respect the phase model differs fundamentally from sequential stage models, such as the model of progressive disillusionment. Recently, Golembiewski *et al.* (1996, p. 2) admitted the confusion caused by the term phase model: 'It is perhaps awkwardly labelled, but that is water long since over the dam'.

Essentially, the categorisation in eight phases boils down to reducing all possible combinations of MBI-scores (over 80 000!) to an eight-point scale. When depersonalisation, reduced personal accomplishment, and emotional exhaustion are assigned weights of 1, 2, and 4, respectively and these weights are added for every phase, an eight-point rating scale emerges ranging from 0 (Phase I) to 7 (Phase VIII) (Burisch, 1989, p. 20).

Research based on the phase model followed two streams. A first stream of research indicated that the incidence of burnout in various phases fairly differs across organisations and countries, with about one-fifth of the employees in the most advanced phase in North America (see Chapter 3). Moreover, burnout tends to cluster in particular work groups: 87% of the employees in the most advanced phase are employed in work groups where more than 50% of their colleagues are assigned to an advanced phase as well (Rountree, 1984). Finally, phase assignments are fairly stable across time: 40% are assigned to the same phase 1 year later, whereas 55% move only one phase up or down (Golembiewski and Boss, 1991). Even after 3 years 30% did not show any change, whereas 43% moved up or down just one phase (Golembiewski et al., 1996; p. 174). Individuals moving to a lower burnout phase reported corresponding improvements in antecedents (i.e. work setting characteristics) and consequences of burnout: they experienced greater job satisfaction and fewer psychosomatic symptoms (Burke and Greenglass, 1991).

The bulk of the studies, however, are included in the second stream that attempts to validate the very notion of progressive phases of burnout. All in all, these attempts have been quite successful (for reviews see: Golembiewski et al., 1986; Golembiewski and Munzenrider, 1988; Golembiewski et al., 1996). Golembiewski and his colleagues claim that over 400 (sic.) variables have been related to their phase model, most of them successfully. That is, employees (or work groups) in more advanced phases exhibit more unfavourable scores on various aspects including: job characteristics (e.g. meaningfulness of work, knowledge of results, workload, autonomy, role clarity, group cohesion, support from supervisors and colleagues); health-related outcomes such as physical symptoms (e.g. cardiovascular complaints, sleeplessness), mental symptoms (e.g. anxiety, depression) and medical consumption (e.g. health care claims, costs of medical treatment); productivity (e.g. headquarters' ratings of management and quality, clients' criticisms, work group effectiveness, overall output achieved); performance (e.g. supervisor's performance appraisal, self-performance appraisal). It must be noted, however, that relationships with hard archival data (e.g. productivity, medical costs, turnover) and biomedical data (e.g. blood chemistry, body mass) are less convincing and robust than with self-reports. Relationships between the variables in the 'environmental model' are depicted in Figure 5.10.

Virtually all relationships in Figure 5.10 are in the expected direction: poor job characteristics are substantially positively associated with burnout and somewhat more weakly with physical symptoms and poor performance. Moreover, burnout is fairly strongly positively related to physical symptoms and poor productivity, but somewhat less strongly to poor performance. Not surprisingly, performance and productivity are positively related to one another, whereas physical symptoms and productivity are negatively interrelated. However, the positive relationship between physical symptoms and productivity is somewhat puzzling: the less healthy the employee feels, the higher the productivity of the work unit. Golembiewski speculates that overtime work may be responsible for producing this counterintuitive result since it may boost productivity at the cost of health complaints.

It must be noted that Figure 5.10 merely summarises the results of associations between pairs of (sets of) variables: the complete environmental model has not been tested. Moreover, the associations depicted in Figure 5.10 do not imply causal relationships. For instance, the fact that employees in more advanced phases of burnout experience higher workload does not mean that high workload leads to burnout, the reverse could also be true.

The work of Golembiewski and his team has been severely criticised on methodological grounds (Leiter, 1993). First, the claim that the content as well as the structure of their modified MBI is congruent with the original version is questioned. For instance, in order to be used outside the human services, the term 'recipients' was replaced by 'co-workers', which changes the interpretation of the depersonalisation dimension dramatically. Furthermore, the progressive phases are heavily biased towards emotional exhaustion. Since this dimension of burnout has consistently been found to be more strongly correlated with environmental stressors and health outcomes: '. . . the relationships of outcome measures or environmental conditions with these phases do not reflect anything about progressive virulence of burnout that was not apparent from the correlations of such measures with burnout' (Leiter, 1993, p. 241). Leiter substantiates his argument by showing that the correlation with phase number is very high for emotional exhaustion ($r = .81$) and drops considerably for depersonalisation ($r = .62$) and reduced personal accomplishment ($r = .42$), respectively.

Despite these criticisms, Golembiewski and his team made clear that burnout is an intrinsic part of organisational life. Not only because about one in every five North American employees is classified in the most advanced burnout phase, but also because burnout is associated with a host of poor job characteristics, and, last but not least, because burnout seems to have severe negative consequences for the organisation.

Burnout as a mismatch between person and job

Recently, Maslach and Leiter (1997) expanded their views on burnout in three ways. First, as opposed to their initial interpersonal approach they no longer consider burnout to be a phenomenon that exclusively occurs in the human services. In their current view: 'Burnout is the index of the dislocation between what people are and what they have to do. It represents an erosion in values, dignity, spirit, and will – an erosion of the human soul. It is a malady that spreads gradually and continuously over time, putting people into a downward spiral from which it is hard to recover' (Maslach and Leiter, 1997, p. 17). Essentially, burnout results from a situation of chronic imbalance in which the job demands more than the employee can give and provides less than he or she needs. This mismatch between person and job is independent from the specific content of the job. It may occur in the human services as well as outside this occupational field. Emotional overload resulting from working with recipients, considered the root cause of burnout in their previous model (Maslach, 1993; Leiter, 1993), is now regarded as a particular aspect of the person-job mismatch.

Secondly, and related to the previous issue, the three original burnout dimensions are now defined in somewhat more general terms: exhaustion refers to feeling overextended both emotionally as well as physically; cynicism refers to a cold, distant, indifferent attitude toward one's work; ineffectiveness refers to a sense of professional inadequacy and the loss of confidence in one's ability. Unlike in the original definition of burnout, these dimensions do not specifically pertain to working with people. As we have seen in Chapter 3, the MBI-GS captures these somewhat broader dimensions of burnout. Besides, in their most recent approach, Maslach and Leiter (1997) considered burnout to be located at one end of the continuum opposite to engagement, consisting of energy, involvement and efficacy. When burnout begins, the employee's sense of engagement begins to fade and a corresponding shift occurs from positive feelings to their negative counterparts. In other words, the original continuum that only had a negative pole and that ran from not burned-out to burned-out is now replaced by a full continuum with two poles: engagement and burnout.

Third, instead of a single root cause for burnout (i.e. emotional overload), six types of person-job mismatches are considered to be potential sources of burnout: work overload (i.e. having to do too much in too little time with too few resources); lack of control (i.e. no opportunities to make choices and decisions, use one's abilities to think and solve problems); lack of reward (i.e. inadequate monetary rewards as well as internal rewards such as recognition appreciation); lack of community (i.e. a loose and unsupportive social fabric, social isolation and chronic and unresolved problems); lack of fairness (i.e. employees are inequitably treated and respect and self-worth are not confirmed); value conflict (i.e. the requirements of the job do not agree with personal principles). Maslach and Leiter (1997) argue that these six person-job mismatches are pervasive in modern organisational life. They illustrate these mismatches with case materials and use their approach to put burn-out research in perspective.

Conclusion

Despite large differences the three approaches agree that similar organisational factors (e.g. qualitative and quantitative job demands, lack of autonomy or control, lack of rewards, incongruent institutional goals or values, and lack of social support or community) are important correlates of burnout. Moreover, they point to the fact that burnout has not only negative effects for the individual but that it is also detrimental for the organisation in terms of lowered productivity and efficiency, and poor quality of service.

5.4 SOCIETAL APPROACHES

Generally, the previous approaches conceptualised burnout as a subjective phenomenon in which perceived stressors tend to be more important than actual environmental conditions. Moreover, burnout was considered to be the product of a poor person-job fit: that is, a disturbed interaction between individual needs and resources, and the various interpersonal or organisational demands, constraints and facilitators. However, the dissemination of burnout in the human services suggests that it may also be a symptom of broader social concerns that goes beyond the professional's subjective experience as well as beyond the particular organisational environment. Hence, it is likely that more deeply rooted structural factors at the societal and cultural level also contribute to burnout, or at least they have set the stage for its appearance. Traditionally, such a sociological perspective has been neglected in the burnout literature as it is dominated by psychology. This section briefly discusses three sociological approaches to burnout that de-emphasise its subjective nature by pointing to specific societal or cultural factors.

Burnout as alienation

Following traditional Marxist thought, Karger (1981) criticised the burnout literature first and foremost for 'privatising' the nature of the problem. Instead of an objective organisational problem, burnout is considered to be a personal and thus subjective problem. The objective realities of work are not taken into account, instead burnout is defined as an individually-based, affect-laden experience that is caused by subjectively perceived stressors.

According to Karger (1981) a similarity exists between burnout and industrial alienation as described by Karl Marx (1818–1883) over one century ago:

> . . . the work he performs is extraneous to the worker, that is, it is not personal to him, is not part of his nature. . . . [he] feels miserable rather than content, cannot freely develop his physical and mental powers, but instead becomes physically exhausted and mentally debased . . . at work he experiences himself as a stranger . . . the relationship of the worker to his own activity as something alien, and not belonging to him. . . . The awareness which man should have of his relationship to the rest of mankind is reduced to a state of detachment in which he and his fellows become simply unfeeling objects. (Marx, 1844, cited in Karger, 1981, p. 275)

Indeed, this description of alienation sounds remarkably like current descriptions of burnout, including exhaustion and depersonalisation.

Put in Marxist terms, burnout results from the 'objectification' of the professional's means of production. That is, the professional's social and interaction skills have become a market commodity. The transformation, in capitalist society, of these skills into merely a means of production results in the distancing of the professional from the recipient. In cases of industrial alienation this objectification occurs in the relationship between worker and inanimate object (e.g. a car, a tv-set, or a computer). By contrast, in the human services, objectification takes place between person and person and is thus likely to foster depersonalisation. This objectification, which is a necessary consequence of the emergence of capitalism that turns more and more genuine skills into market commodities, deprives people of the source of personal identity they derive from their work. In other words, their work loses personal meaning when it becomes a mere product for the market. In its turn, objectification leads to other negative consequences that might foster burnout, such as fragmentation of work, competition among colleagues, and loss of autonomy. Indeed, specialisation, poor social climate at work, and bureaucratic interference have been associated with burnout (see Chapter 4).

In conclusion, particular societal developments, most notably objectification, contribute to the development of burnout through a process of alienation. It is crucial to note that objectification and its consequences are considered to be objective social conditions that exist independently from the professional's subjective perception thereof.

Burnout as a discrepancy between surface and latent functions of organisations

Handy (1988) criticised the occupational stress and burnout literature for using too simplistic a notion of person-environment interaction: the perceived imbalance between organisational demands and individual response capability. Because this view is strictly psychological it neglects the social dynamic that underlies the complex transactions between employee and organisation, she argues. Besides, this interaction takes place within a broader social, cultural and historical context.

According to Handy (1988), psychologically-based models of burnout can be augmented by including the sociological notion of manifest versus latent functions of organisations and the corresponding surface and deep structures. She presents an example from the educational field to clarify this distinction:

> The manifest function of schools is to offer all children equal educational opportunities. However, one of their latent functions is to help socialise children into a class society and prepare them to take their place in work organisations. This function frequently contradicts the educational system's stated aims since it involves facilitating the reproduction of specific forms of social inequality and, in consequence, it may not fully be recognised by either staff or pupils. (p. 355)

The manifest function of schools is apparent in the surface structure (e.g. curriculum, building, exams), whereas their latent function is less explicit and forms part of the deep structure of the educational system. Furthermore, according to Handy (1988), the prevailing simplistic view on employee-organisation interaction has focused almost exclusively on the organisation's structures and functions and has therefore severely undervalued the proactive role of employees. Employees are not merely passive victims, they are active agents who tend to create their own social reality that may or may not differ from the formal aims of the organisation. This implies that rather than a fixed 'giving' organisational reality it is a negotiated social order in which issues of power and conflict play a crucial role.

The relevance of these theoretical notions for burnout is illustrated by two empirical studies. In her study on stress in psychiatric nursing Handy (1991) argues that psychiatry's 'dual mandate' to control and to care for the mentally ill creates fundamental contradictions in the daily work of psychiatric nurses. This dual mandate expresses psychiatry's manifest function (i.e. to cure patients) as well as its latent function (i.e. to control social deviants). Handy (1991) found that the nurses' daily activities centred on social control and on the maintenance of ward routines and that such activities were often incompatible with therapeutic strategies. The daily practice of social control ran counter to the nurses' self-image as professional carers as well as to their therapeutic ideals. The awareness of this discrepancy sometimes led young nurses to piecemeal attempts to develop more therapeutically-oriented relationships with individual patients. However, such innovations often failed because of the nurses' inexperience and because of the control-oriented hospital structure. Repeated failure of their attempts not only triggered feelings of incompetence in these young nurses, but also led them to blame their patients for being unmotivated. Older nurses had adopted a more instrumental and routine-oriented attitude toward work, partly in order to serve as a role model for their younger colleagues. By doing so, the potential for change that was present in younger nurses was channelled into the maintenance of existing patterns.

In a somewhat similar vein, Satyamurti (1981) argued that the stresses experienced by social workers have their roots in the imbalance between the demands which the state places on them and the resources it places at their disposal. Social workers have to solve their clients' practical and emotional problems which can mostly be tracked back to chronic material deprivation. Unfortunately, social workers have limited access to the material resources necessary to alleviate the root causes of their clients' problems. As a consequence, social workers often fail, they feel exhausted and inadequate. Because of their individual focus, they often, wrongly, blame their clients for the negative outcomes of their helping strategies. Thus, as a result of personal feelings of stress generated by structural contradictions of their work, many social workers ended up blaming the very people that they had originally intended to help. Interestingly, Satyamurti (1981) also found that those social workers who showed a clear understanding of these contradictions felt less incompetent and did not blame their clients.

In summary, societally induced structural contradictions in human services organisations may contribute to burnout. More specifically, burnout may be caused by the discrepancies between the manifest and latent functions, and the surface and deep structures of organisations that profoundly influence the actions and understandings of individual employees.

Burnout as a cultural product

The purpose of Meyerson's (1994) approach was to illuminate the social construction of burnout. More specifically, she was interested in the way the professionals' interpretation of burnout reflects and reinforces particular institutional systems. Accordingly, her approach stressed human agency: institutional systems not only regulate the behaviour of their members, they are also reproduced in their actions and interpretations.

Meyerson (1994) studied two institutional systems or cultures that are encountered among social workers in hospitals: the medical model and the psychosocial model. The former is characterised by: the construction of that which is not statistically normal as 'disease'; the belief that the individual is the locus of disease; the belief in order, clarity and control. In contrast, the psychosocial model is characterised by: an acceptance of a multiple notion of 'normality'; the belief that clients' conditions have social (rather than individual) roots; a belief in the self-determination of clients.

She found that in four of the five hospitals she examined, the medical model surfaced as the dominant institutional system or culture that governed how social workers behave and interpret their own and others' behaviours. In the remaining chronic, non-teaching, rehabilitation hospital the alternative psychosocial model of health and illness prevailed as the dominant institutional system. As expected, burnout was interpreted differently depending on the dominance of the particular culture. When the medical model prevailed, burnout was considered a pathological condition that had to be controlled, a disease that one caught and tried to cure. Professionals and supervisors in those settings were very reluctant to admit that burnout was a problem among social workers. By contrast, when the psychosocial model prevailed burnout was viewed as a social phenomenon and a normal part of work. In this psychosocial culture, burnout did not represent a personal or professional failure, but an unavoidable experience and a normal part of the job. Interestingly, burnout was encountered much more frequently in the chronic hospital with the psychosocial culture than in the other hospitals in which the medical model was clearly dominant. This illustrates the self-perpetuating nature of the process: the presence (or absence) of burnout reinforces the very culture that produced (or did not produce) it.

Thus, variation in the amount of burnout across organisations may be due to different assumptions about its nature and differences about the legitimacy of expressing burnout symptoms. Such cultural differences tend to be self-perpetuating.

Conclusion

Societal approaches to burnout highlight the role of objective structural and cultural determinants that exist irrespective of the individuals' subjective interpretation of reality. Another common feature is their transactional or dialectical nature; workers are not considered to be passive victims but are active agents who, individually and collectively, shape their own working and living conditions that may or may not contribute to burnout.

5.5 SUMMARY: FROM KALEIDOSCOPE TO SYNTHESIS

After having read the previous nineteen approaches to burnout the reader may by now feel somewhat confused by the emerging kaleidoscopic picture. This should not come as a surprise, though. We noted in the introduction to this chapter that no overarching theory of burnout exists, and there probably never will be one. In order to explain burnout complex interactions between numerous aspects, four different levels (i.e. individual, interpersonal, organisational, and societal) have to be taken into account. It is evident that no single theoretical approach can accomplish this and account for the huge complexity. Hence, every theoretical explanation of burnout necessarily remains a fragmented, piecemeal explanation. Nevertheless, in this concluding section we make an attempt to identify common ground that is covered by several approaches. Building on these similarities, we propose an integrative and descriptive model of burnout. This rather abstract model is not a psychological theory that explains burnout, rather it is a heuristic model that schematically summarises common issues that are included in the previous approaches.

An integrative model of burnout

Three recurrent themes run through many approaches: a strong initial motivation is a necessary condition for developing burnout; burnout is associated with an unfavourable job environment; the burnout process is self-perpetuating because of the use of inadequate coping strategies.

Strong motivation

The crucial role of a strong initial motivation in the development of burnout is widely recognised. Many approaches emphasise the importance of highly stacked goals, expectations, intentions, and aspirations and of a strong sense of enthusiasm, idealism, involvement and commitment. This is not only a common element in virtually all individual approaches, but also in most interpersonal approaches: motivation to help in the social competence approach; detached concern in the emotional overload approach; restoration of reciprocity in the social exchange approach; emotional labour in the corresponding approach. In addition, it is found in the organisational reality shock approach (career orientation) and in the person-job mismatch approach (initial sense of engagement). Obviously, most approaches agree that a strong, initial motivation is a necessary condition for developing burnout. It seems that Pines (1993, p. 41) was right: 'In order to burn out, one first has to be "on fire"'.

Usually, in the human services, professionals are strongly motivated to help their recipient. A strong motivation to help is a crucial asset and a key to success. Without being dedicated, committed, and involved the professional's goals are difficult to obtain. Hence, a paradox exists: the most valuable and successful professionals are those who, for that very reason, run the largest risk of burning out.

Unfavourable job environment

Many approaches assume that burnout is likely to result when the professional's strong motivation (to help) stands in sharp contrast to his or her everyday experience on the job. In other words, a person-job fit rationale underlies many approaches:

the highly motivated professional gradually becomes frustrated and feels exhausted because intentions do not match reality. This line of reasoning is found most explicitly in the person-job mismatch approach (Maslach and Leiter, 1997), but it is also encountered in early descriptive accounts of burnout (Freudenberger and Richelson, 1980; Edelwich and Brodsky, 1980), as well as in more theoretical analyses using cognitive learning theory (Meier, 1983), action theory (Burisch, 1989), psychodynamic theory (Fischer, 1983; Garden, 1991), social exchange theory (Buunk and Schaufeli, 1993), and in approaches that stress the role of emotional labour (Morris and Feldman, 1996; Ashforth and Humphry, 1993), emotional demands (Maslach, 1982a; 1982b; Leiter, 1993), and, last but not least, reality shock (Cherniss, 1980a; 1980b).

A person-job misfit is particularly likely to occur in the human services because it seems that, on the one hand, the human services attract highly motivated individuals (feeling types), whereas on the other hand working in these services can be a particularly frustrating experience. For at least four reasons, the job environment in the human services is unfavourable and thus potentially stressful. First, emotional demands are high because human services professionals work with difficult and troubled recipients who take the professionals' efforts for granted (lack of reciprocity). Second, because of the nature of the job the action pattern is likely to be disturbed (e.g. professionals receive late, poor, or incomplete feedback, or goals, criteria, and objectives are unclear). Third, the organisational environment is usually not very supportive (e.g. resources are chronically lacking). Finally, societal and cultural pressures are high so that conflicts are likely to result (e.g. conflicting demands from society's dual mandate to simultaneously care and control).

As a result, another paradox exists: the human services are crucially dependent upon the professional's motivation to help, yet the job environment is inherently unfavourable so that the professional's motivation is bound to be frustrated. In other words, the human services offer an unfavourable job environment with structurally built-in sources of frustration that potentially foster burnout.

Inadequate coping strategies

Generally, the dynamic, self-perpetuating nature of burnout is emphasised in one way or another. It is assumed that a positive or a negative feedback loop exists depending on the way the professional copes with the distress that results from the mismatch between strong motivation and unfavourable job environment. Such feedback loops are graphically illustrated in Figures 5.2, 5.3, 5.4, 5.5 and 5.9, and are also found in the approaches of Freudenberger (1983), Fischer (1983), and Ashforth and Humphry (1993). When a successful coping strategy is followed (e.g. active problem solving, detached concern) goals are achieved, professional efficacy is enhanced, and a sense of existential significance is fostered. Some authors consider professional efficacy to be the opposite of burnout that leads to positive individual (e.g. health, self-esteem) and organisational (e.g. productivity, good quality of services) outcomes (Cherniss, 1993; Leiter, 1993; Harrison, 1983; Maslach and Leiter,

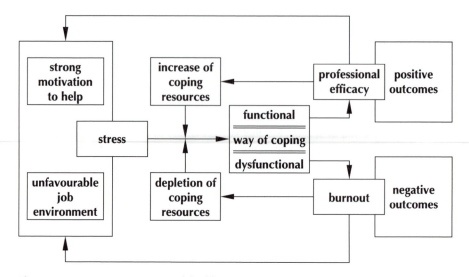

Figure 5.11   An integrative model of burnout

1997). By contrast, when a poor coping strategy is adopted (e.g. physical or mental withdrawal, misattribution) burnout is likely to develop, including the negative individual (e.g. ill-health, depression, low self-esteem) and organisational (e.g. absenteeism, poor performance) consequences. Depending on the way burnout is conceptualised it may include some aspects of inappropriate coping behaviours in the sense of mental distancing (e.g. depersonalisation).

Hence, burnout is a self-perpetuating process not only because it impedes the attainment of professional goals but also because it depletes coping resources – 'spiral of loss' (see Figure 5.11). Likewise, but from a positive stance, professional efficacy increases when goals are attained and coping resources are built up – 'spiral of gain'.

Integration of common ingredients

Figure 5.11 depicts an integrative, sequential-dynamic model of burnout that emerges when the three most common ingredients, previously discussed, are put together.

The model displayed in Figure 5.11 is sequential since it assumes that the distress that results from the discrepancy between strong motivation and unfavourable job environment leads to burnout or its counterpart, professional efficacy, depending on the professional's way of coping with this misfit. In addition, the model is dynamic: it assumes that the processes involved are self-perpetuating. That is, burnout increases distress and depletes coping resources, whereas professional efficacy reduces distress and augments coping resources. In fact, this integrative model is a graphical representation of our working definition of burnout, presented in Chapter 2.

CHAPTER SIX

What to do about it?

Interventions

Since burnout first emerged as a social problem rather than as a scholarly concept, ways of combating and curing it have received special attention from the outset. Traditionally, there is a strong focus on immediate action in the burnout literature. As Maslach (1984, p. 159) has put it: '... numerous solutions are being proposed because of the pressing need to do something about burnout and to do it now'. This chapter provides a comprehensive overview of various burnout interventions that have been recommended in the past decades. We start by proposing a general classification scheme that distinguishes between approaches that are either aimed at the individual, at the individual/organisational interface, or at the organisation. Next, we discuss in greater detail over thirty different interventions. In a separate section, specific burnout prevention workshops that combine various intervention techniques are discussed.

6.1 CLASSIFICATION OF INTERVENTIONS

Basically, interventions to reduce job stress or burnout may be directed at three levels (cf. De Frank and Cooper, 1987):

■ **The individual** By learning to cope with stress better the individual prevents the occurrence of negative psychological effects. Individually-based interventions focus on the person's reactions to stressful circumstances, irrespective of their context.

■ **The individual/organisation interface** By increasing the employee's resistance to specific job stressors their vulnerability decreases. Essentially, approaches at this level are aimed at the interplay of individual and organisation. They focus on the employee – the working individual in the context of the organisation.

■ **The organisation** By changing the work situation through organisational based interventions the source of the problem is tackled. Many interventions at this level are mere 'band aids' for the real problem of job stress and burnout, though. They are primarily designed for increasing productivity, improving quality, or reducing costs.

In addition to the focus or level of the interventions they may serve different purposes. Table 6.1, which classifies interventions that are mentioned in the burnout literature, distinguishes five purposes. Since early detection is crucial for combating burnout its identification is included as a first distinct aim ('forewarned is forearmed'). Usually, prevention strategies are classified as being primary, secondary, or tertiary in nature, each having different purposes: the aim of primary prevention is to reduce risk factors, or to change the nature of the stressors; the aim of secondary prevention is to alter the ways individuals respond to stressors; and the aim of tertiary prevention is to heal those who have been traumatised. However, instead of tertiary prevention we prefer to speak about treatment because a relatively clear distinction exists between preventing burnout among those who are at risk, and treating those who are actually burned-out. Typically, the former are still at work, whereas the latter are on sick-leave and are suffering from severe burnout. Finally, unlike most classifications, we include as a distinct category the rehabilitation of those who have suffered from burnout and returned to work. Please note that by definition rehabilitation goes beyond the purely individual level in the sense that it inevitably takes into account the employee's job. Therefore, the cell in the upper right corner of Table 6.1 is empty. Moreover, since we focus on job-related mental problems (burnout), a general discussion of psychotherapy and counselling is beyond our scope so that the individual treatment cell also remains empty: psychotherapy and counselling are discussed exclusively in the context of work.

Although Table 6.1 may look quite impressive at first glance, few critical notes have to be made because any attempt to force all existing burnout interventions neatly into one scheme is doomed to fail. Interventions simply differ too much, not only with respect to their purpose and focus, as recognised in Table 6.1, but also in terms of target-group and scope. For instance, the distinction between primary and secondary prevention is not always as clear-cut as Table 6.1 suggests, particularly at the interface and organisational levels. Basically, primary prevention is aimed at all employees, whereas secondary prevention is aimed at those who are at risk for burnout. Of course, one can argue that what is good for the average employee is also good for employees who are at risk. So, it is essentially the targeted group that determines whether an intervention is classified as being primary preventive or secondary preventive. Likewise, interventions may differ considerably in scope. For instance relaxation, time-management, and self-monitoring are specific techniques, whereas psychotherapy, organisational development, and corporate fitness programmes are comprehensive activities that usually include various techniques. Relaxation training may be included in psychotherapy or in corporate fitness programmes, and time-management training may be part of an organisational development project. Nevertheless, despite its inevitable imperfection we believe that our classification

Table 6.1 Overview of burnout interventions

	Identification	Primary Prevention	Secondary Prevention	Treatment	Rehabilitation
Focus on Individual	Self-monitoring Self-assessment	Didactic stress management Promoting a healthy lifestyle	Cognitive-behavioural techniques Relaxation		
Focus on Individual/ Organisational Interface	Personal screening	Time-management Interpersonal skills training Promoting a realistic image of the job Balancing work and private life	Peer-support groups Individual peer-support Coaching and consultation Career planning	Specialised counselling Psychotherapy Referral	Individual guidance and assistance Change jobs
Focus on Organisation	Stress audit Psychosocial check-up	Improving the job content and environment Time scheduling Management Development Career management Retraining Corporate fitness and wellness programmes	Anticipatory socialisation Conflict management, communication, and decision-making Organisational Development	Institutionalisation of Occupational Health and Safety Services Employee Assistance Programmes	Outplacement

scheme provides a useful structure for discussing the plethora of burnout interventions in a more systematic fashion.

6.2 INTERVENTIONS PRIMARILY AIMED AT THE INDIVIDUAL

Most individual level interventions are well-established and have a long and successful history in clinical or health psychology. They are rather general in nature since they focus on managing stress per se rather than on combating burnout in particular. Principally, the six individual strategies distinguished in Table 6.1 seek either to increase awareness (i.e. self-monitoring, self-assessment, and didactic stress management) or to reduce negative arousal (i.e. promoting a healthy lifestyle, cognitive-behavioural techniques, and relaxation).

Self-monitoring

The underlying idea of self-monitoring is that, by explicitly focusing on the signs and symptoms of distress, the individual's self-awareness is increased ('know thyself'). Self-awareness in the sense of acknowledging that there is a problem is a necessary first step to tackle that problem. Self-understanding begins with self-observation: through self-monitoring certain patterns may be observed that may serve as a starting point for subsequent steps. For instance, it may become clear that one's headaches occur exclusively prior to particular meetings, or that one feels especially exhausted after teaching a particular class. A powerful self-monitoring technique is to keep a stress-diary, a personal record or log of stress-symptoms and related events (see Box 6.1).

Depending on its format, a stress-diary not only provides information about the frequency and context of particular symptoms, but it also includes related thoughts, feelings, and ways of coping. The diary should be kept for an extended period of

Box 6.1   Example of a stress-diary

Physical stress symptoms	Time of the day	Where? What occasion? Who was involved?	What thoughts or feelings?	What have I done?
Headache, muscle tension in the neck	4 p.m.	At work, during intake of new client	Frustration and anger	Drank cup of coffee
....	....	....	....	....
....	....	....	....	....

Source: Maslach (1982b, pp. 99–100)

time, say several weeks. A study that used a stress-diary with correctional officers showed that, as expected, the number of stressful events at work was positively related to the officer's negative affect at the end of the workday (Peeters *et al.*, 1995).

Self-assessment

Several paper-and-pencil tests are available for self-assessment of burnout (see Chapter 3). Although such tests may increase the individual's awareness of certain burnout symptoms, they do not allow valid conclusions about the level of burnout experienced relative to a comparison group. Only the MBI provides statistically derived cut-off scores that can be used as a standard of comparison: the individual may roughly classify her- or himself as 'low', 'average' or 'high' in burnout compared with workers in such occupational fields as teaching, medicine, social services, or mental health. In Chapter 3 the diagnostic use of the MBI is extensively discussed.

Didactic stress management

Didactic stress management refers to all kinds of information about stress and burnout that are provided with the intention of increasing awareness and improving self-care. Books, newspaper or magazine articles, brochures, flyers, TV-programmes (talk-shows!), films, audio cassettes, lectures, and, recently, internet-sites may all serve this purpose. Typically, didactic stress management is not only concerned with symptoms or causes, but especially with cures and remedies. Information is usually provided about relaxation exercises, nutrition, physical fitness, cognitive techniques, and social skills. A good example is the workbook of Jaffe and Scott (1988) that includes many useful tips, tricks, and exercises that teach the individual how to deal with job stress and burnout (see also Fontana, 1989; and for teachers Gold and Roth, 1993).

   Although sincere attempts to increase the individual's awareness about job stress and burnout have to be welcomed, there exists also a serious drawback: the so-called 'medical students-syndrome'. When confronted with the symptoms of burnout individuals might falsely relate these to themselves and thus assume that they themselves actually suffer from it. This phenomenon was first observed among medical students who were said to suffer from those particular physical symptoms that had been described vividly in the previous lecture. In order to avoid this syndrome, didactic stress management should be balanced and not only include weaknesses (e.g. symptoms) but also strengths (e.g. coping resources). The above-mentioned books provide good examples of such a balanced approach.

Promoting a healthy lifestyle (physical exercise)

The comment of Homer: '*mens sana in corpore sano*' ('a healthy mind in a healthy body') applies to the prevention of burnout as well, at least according to Maslach (1984). She emphasised that employees should not only take care of their minds, but also of their bodies because physical well-being is an integral part of emotional well-being and may make an individual more resistant to job stress. A healthy life-style includes regular physical exercise, proper nutrition, weight control, no smoking, enough sleep, and periods of rest for relaxation and recharge during the workday and thereafter.

Of these approaches, physical exercise is perhaps the most powerful antidote to stress. Research has shown that routine vigorous activity is an effective strategy for preventing the negative effects of stress such as depressed mood, depression and anxiety (McDonald and Hodgdon, 1991). The type of physical activity that produces the most positive effect is aerobic exercise: jogging, running, cycling, and swimming. In order to be effective, the exercise should be carried out three to four times a week over 30–40 minutes at 50–60% of maximal working capacity (Ross and Altmaier, 1994). So jogging for half an hour once a week, or playing a tennis-match every fortnight will not do! The exact process by which physical exercise influences the individual's psychological state is not fully understood yet. It is suggested that physical exercise prevents the negative effects of stress and thus reduces burnout by lowering arousal.

Cognitive-behavioural techniques

The ancient Greek philosopher Epictetus, one of the founding fathers of stoicism, argued that men are not disquieted by things themselves, but by their idea of things. In other words, individuals do not respond directly to their environment, instead they respond to their own interpretation of that environment. Emotional responses (e.g. anger, frustration, fear, depression, anxiety) are not provoked by the situation, but by the cognitive label attached to it. Furthermore, cognitive-behavioural techniques assume that cognitions (thoughts), emotions (feelings), and behaviours (actions) are causally interrelated in the sense that:

thoughts → feelings → actions

Following this logic, changing thoughts so that the situation is 'appraised' differently will reduce negative feelings and ultimately eliminate undesirable behaviour. This line of reasoning provides the basis for cognitive-behavioural techniques for stress and burnout prevention. These techniques imply that instead of the things themselves, the individual's ideas about those things are changed. In the previous chapter we saw that particular cognitions, such as high expectations, are likely to play a role in the development of burnout. Clinicians have often observed the existence of a vicious circle in burnout victims that centres on faulty or irrational ideas that are summarised in Box 6.2.

For instance, a teacher holds strong irrational beliefs about her own professional competence (3. in Box 6.2). Suppose, for one reason or another, she fails to achieve her highly stacked goals. As a result, her idea about her own competence is challenged and she starts feeling tense and depressed, which impairs the quality of her teaching and thus reduces the likelihood of achieving her goals in the near future.

Box 6.2   Ten common irrational ideas in the human services

1. the idea that it is a dire necessity for a helping professional to be loved or appreciated by every client

2. the idea that one must always enjoy the favour of one's supervisor

3. the idea that one must be thoroughly competent and successful in doing one's job if one is to consider oneself worthwhile

4. the idea that anyone who disagrees with one's own ideas and methods is bad, wicked, or villainous and therefore becomes an opponent to be scorned, rejected, or blamed

5. the idea that one should become very upset over one's recipients' problems and failings

6. the idea that it is awful and catastrophic when things are not as recipients and the institution would like them to be

7. the idea that unhappiness is caused by recipients or the institution and that one has no control over one's feelings

8. the idea that until recipients and the institution straighten themselves out and do what is right one has no responsibility to do what is right oneself

9. the idea that there is a right, precise, and perfect solution to human problems and that it is catastrophic if that solution is not found

10. the idea that dangerous and fearsome things can happen to recipients, which are cause for great concern and should be continuously dwelled upon

Adapted from: Edelwich and Brodsky (1980, p. 206)

This, in its turn, will further strengthen her doubts about her competence as a teacher, and so on. Thus, the circle is closed.

Several specific cognitive-behavioural techniques are available to break that vicious circle. For instance, **cognitive appraisal** teaches the individual to assess the severity of a stressor by taking into account the real perspective of one's stressful situation. That is, individuals are encouraged to ask themselves certain questions systematically when a negative event occurs. For instance, 'What is the worst possible outcome of missing this particular deadline?' (6. in Box 6.2). By answering this kind of question the individual learns that missing a deadline neither threatens his life, nor his future job security. By asking, in addition, about the positive effects of having missed the deadline (e.g. being allowed more time for a client who needed particular attention), a new perspective may be gained. Essentially, a less negative and more realistic framing of a stressful event is envisaged by cognitive reappraisal. Therefore, this technique is also called perspective taking. Edelwich and Brodsky (1980) point to the fact that unrealistic expectations often lie at the core of burnout and that therefore alternative perspectives are crucial for its prevention. They recommend setting specific and realistic goals, focusing on the successes and not on the failures, setting long-term as well as short-term goals, focusing on the process instead of the result, and not misinterpreting results self-referentially (i.e. do not blame yourself).

A somewhat related technique is **cognitive rehearsal** that involves helping individuals tolerate stressors by anticipating them before they happen. This is achieved by visualising a potentially stressful event *in vitro* and by practising or rehearsing how to respond. For instance, an employee finds it extremely difficult to say 'no' to the supervisor. Vividly imagining this event and rehearsing how to respond (e.g. decisively but presenting a rationale and suggesting an alternative) prevents future stress reactions in such situations. It is important that rehearsal occurs when the individual is relaxed, so that this technique is usually combined with teaching relaxation skills.

Probably the most popular technique is cognitive restructuring, the core element of **Rational Emotive Therapy** (RET) (Ellis, 1962). This approach is based on the assumption that irrational thoughts or beliefs lead to stress and, hence, that restructuring these cognitions reduces stress. RET can be described using the so-called A-B-C-D-E framework:

- Activating experience: an event happens,
- Belief: an irrational belief about A exists,
- Consequence: this belief leads to an emotional or behavioural consequence,
- Disputing, Debating, Discriminating, and Defining are questioning and challenging techniques that are used to overcome irrational beliefs,
- Effect: the individual acquires a new philosophy which helps him or her to think more rationally and constructively.

Box 6.3 presents an example of how this step-by-step procedure can be used for analysing and redirecting an individual's thinking. RET is not only very popular but is also a quite powerful approach to preventing burnout because irrational beliefs are widespread among human services professionals (for common examples see Box 6.2).

Recently, Malkinson *et al.* (1997) showed that RET was effective in reducing burnout among blue-collar female workers. Compared with a non-treated control group, those who received RET-training during six meetings of 3 hours each showed lower burnout levels immediately after the training as well as at the 1-year follow-up.

Finally, an often used approach that combines some of the techniques outlined above is **Stress Inoculation Training** (SIT) (Meichenbaum, 1985). In addition to altering the way the individual processes information about a stressful situation, SIT identifies and teaches cognitive and behavioural coping skills to change unproductive ways of reacting in three stages:

Box 6.3   Rational Emotive Therapy

Activating experience: A man whom one is counselling has started beating his wife again.

Belief: 'It's my fault. If I were a good counsellor, he wouldn't be beating his wife. If I can't stop him from doing it, my effectiveness as a counsellor will be called into question.'

Consequences: One loses a night's sleep worrying about the case and one begins to feel run down.

Disputing: 'What makes me think that I have the power to stop someone from doing what I don't want him to do? Why should I be able to prevent bad things from happening? Why does one setback make me a bad counsellor? And why does doing a bad thing make this man a bad person?'

Effect: 'Even if I were the best counsellor in the world, I couldn't count on being able to stop this man from beating his wife. He has been and is subjected to many influences on his behaviour besides myself. He is doing something bad now, but that does not mean that he will always do it. It makes him a fallible, not a bad, person. Because he's doing something bad, though, it is my responsibility to do everything that is reasonably within my power to stop him.'

Source: Edelwich and Brodsky (1980, p. 207)

- **Educational phase** Identify the problem and provide a conceptual framework for understanding one's specific response to stress.

- **Rehearsal phase** Teach the individual a variety of ways to cope with the stressful situation.

- **Application training** Gradual exposure to the stressful situation and application of the newly developed skills.

Whereas RET tends to focus on the presence of maladaptive cognitions or irrational beliefs, SIT focuses on the absence of adaptive cognitive-behavioural responses. Two studies that used SIT have demonstrated positive results with burnout. Freedy and Hobfoll (1994) enhanced nurses' coping skills by teaching them how to use their social support and individual mastery resources and found a significant reduction in emotional exhaustion and depression in the experimental group compared with the non-treated control group. West et al. (1984) taught nurses several coping skills simultaneously in the rehearsal phase (i.e. relaxation, assertiveness, cognitive restructuring, and time-management) and observed at the 4 month follow-up a reduction in emotional exhaustion and an increase in personal accomplishment as well as a decrease in anxiety and systolic blood pressure. More detailed analysis revealed that the rehearsal phase was the principal ingredient of SIT.

Relaxation

Since burnout is characterised by a high level of arousal, an inability to relax is often observed among those who suffer from it. This inability perpetuates the vicious circle of exhaustion: feeling fatigued but not being able to relax aggravates one's arousal, making it even more difficult to relax, thereby further depleting one's resources, and so on. Directly rushing from one's work into leisure activities has a similar effect, particularly if these activities are pursued in an achievement-like and competitive manner. Instead of relaxation, leisure becomes stressful in itself.

Relaxation is considered to be a universal antidote to stress. Therefore, it is the cornerstone of virtually every stress-management programme. According to Murphy (1996), roughly 75% of these programmes include relaxation, often in combination with cognitive-behavioural techniques. The goal of relaxation is to teach the aroused individual how to produce voluntarily a positive, alternative physiological response, a state in which (s)he deliberately eliminates the undesirable physiological effects of stress. Whereas the stress-reaction increases respiration rate, heart rate, blood pressure, and muscle tension, relaxation seeks to decrease it. And whereas the stress-reaction decreases the galvanic skin response, and the alpha and theta brain waves, relaxation seeks their increase. In addition to these physiological effects, it is claimed that relaxation reduces psychological symptoms such as anxiety, depression, burnout, and job dissatisfaction. Generally speaking, the psychological effects of relaxation are more conclusive and consistent than are the physiological effects (Murphy, 1996).

In Box 6.4 the four most popular relaxation methods are described.

It is important to note that all relaxation techniques are self-control strategies which, by definition, depend upon the individual's active participation. Relaxation skills have to be learned, which is not easy for aroused and anxious individuals, particularly because initially relaxation often creates unusual and frightening feelings such as dizziness, loss of control, ringing in the ears, and sudden muscle contractions. It follows that the more burned-out an individual is, the more (s)he needs active, outside support. Furthermore, despite its preventive function, no relaxation method is intended as a form of treatment in and of itself: relaxation should be integrated in an overall programme.

Higgins (1986) compared two individual approaches to reduce burnout among women from various helping professions: relaxation training through Progressive Muscle Relaxation; cognitive and behavioural skills training (i.e. time-management, assertiveness training, and RET). At the post-test after seven sessions, levels of emotional exhaustion had decreased equally and significantly in both experimental conditions, whereas no significant changes were observed in two non-treated control groups. Obviously, both programmes were equally effective.

Box 6.4   Relaxation techniques

Muscle relaxation The best known technique in this category is *Progressive Muscle Relaxation* in which each of fifteen different muscle groups is tensed and then subsequently relaxed in isolation of other muscle groups. Although a full training programme requires six to ten sessions, Maslach (1982a; p. 151) presents an *Instant Deep Muscle Relaxation Drill* that can be practised during regular breaks or shortly before a stressful event occurs.

Deep breathing is a very simple, yet useful, technique. The individual is instructed to breathe from the abdomen, thus avoiding shallow breaths that are associated with the stress response.

Meditation requires four essential components: (1) a quiet place; (2) a comfortable position; (3) an object, sound, feeling, or thought to dwell upon ('*mantra*' – Sanskrit for 'sacred counsel'); (4) a passive attitude. Examples of meditative approaches are: yoga, transcendental meditation, guided imagery, self-hypnosis and autogenic suggestion.

Biofeedback uses equipment to reveal to individuals some of their internal physiological events in the form of visual or auditory signals in order to teach them how to manipulate these otherwise unnoticed events by manipulating the displayed signals. In other words, signals about biological events are fed back. Usually, these events involve myocardial activity (heart beat), muscle tension, and brain wave activity. Initially, biofeedback was viewed as a panacea but it fell somewhat from grace in the 1980s when more rigorous and controlled studies were available which suggested that it was effective only for particular problems such as tension headaches, migraine, and stress-related hypertension.

6.3 INTERVENTIONS PRIMARILY AIMED AT THE INDIVIDUAL/ ORGANISATIONAL INTERFACE

Since burnout is defined as a work-related phenomenon, it is not surprising that most interventions deal with organisational issues, either directly or indirectly, by focusing on the interaction between employee and organisation. We will discuss 15 interventions that are aimed at the individual/organisational interface. They seek to: increase awareness (i.e. personal screening); improve individual coping skills (i.e. time-management, interpersonal skills training, promoting a realistic image of the job, and balancing work and private life); provide emotional and instrumental support at work (i.e. peer-support, coaching, and career planning); cure target complaints by intensive treatment (i.e. psychotherapy, counselling, and referral); and rehabilitate burned-out employees (i.e. individual guidance and assistance, and changing jobs).

Personal screening

When employed in combination with a burnout instrument, psychosocial screening devices may be used to assess the employee's exposure to work stressors and their relation with burnout. Typically, such screening instruments as the Occupational Stress Indicator (OSI – Cooper *et al.*, 1988) include several job stressors, ways of coping with stress, and mental and physical stress-reactions. Personal screening not only provides the employee with his or her level of burnout relative to that of other members of the organisation or of the professional group, but it also presents a more comprehensive picture that includes particular sources of stress, other stress responses, and personal ways of dealing with stress. By contrast with the stress audit (see later), which is conducted and analysed at the organisational level, personal screening is an individual matter. For instance, under the Dutch Working Conditions Act, employees are entitled to ask for a personal screening to be performed by the Occupational Health and Safety Service.

Time-management

Employees who are at risk of burning out generally feel that they have to do too much in too little time. They experience time pressure and feel that they are not able to perform their core tasks at an appropriate level. Rather than an individual problem, time pressure is a pervasive structural problem in many organisations. Yet, the individual employee has to cope with it by efficiently using his or her time. Maslach (1982b; pp. 89–95) suggested several strategies that might prevent burnout, such as 'working smarter instead of harder', 'breaking away', and 'taking a brief time-out'. These strategies serve as emotional breathers that allow the employee to relax and to gain some psychological distance from his or her work. Unfortunately, in many cases such time-management strategies do not work since employees lack the appropriate skills so that a formal training that usually focuses on three areas is helpful:

- **Knowledge acquisition** The employee's job responsibilities, duties and authority are clarified. By keeping a log of time spent on specific duties, the employee is able to see if the actual allocation of time matches the duties and responsibilities of the job.

- **Prioritising** In addition to priorities at work, employees have needs and aspirations outside work and beyond their current positions that also have to be considered.

- **Identify 'time robbers'** such as meetings, visitors, returning phone calls. Employees are encouraged to conserve time (e.g. speed reading), control time (e.g. realistic planning), and to make time (e.g. effective delegation).

Time-management training is considered to be a key coping resource that has been successfully included in comprehensive stress inoculation programmes to reduce burnout (Higgins, 1986; West *et al.*, 1984).

Interpersonal skills training

Typically, in most jobs, professional skills are considered to be more important than interpersonal skills. For instance, in the human services interpersonal skills are taken for granted and not recognised as a necessary asset for professionals. Yet, these skills are crucial not only because human services professionals work with recipients, but also because they collaborate in teams. Demanding interpersonal relations with recipients as well as with co-workers and supervisors play an important role in the development of burnout (see Chapter 4). Poor relations with recipients increase the professional's qualitative workload, whereas difficult relations with others at work undermine the potentially beneficial role of social support.

Maslach (1982b, pp. 136–141) specifies several essential interpersonal helping skills such as: 'how to start, stop and keep things going' (i.e. how to manage the various stages of the helping process by 'breaking the ice', dealing with non-compliance, and so forth); 'how to deal with different people' (i.e. people who vary according to their gender, race, age, cultural background, personality, values, and attitudes); 'how to talk about unpopular topics' (e.g. ask tough or embarrassing questions, discuss sensitive issues, or deliver bad news). Because of the increased entitlement of recipients, professionals have to deal more and more with aggressive behaviours. Therefore, Abernethy (1995) developed an anger management training programme for law enforcement personnel that teaches them how to cope effectively with annoying inmates who provoke feelings of anger and hostility.

In addition, many professionals lack assertiveness. That is, the ability to respond in a straightforward manner with regard to what one believes, feels and wants. In the human services, the characteristic idealistic, self-sacrificing attitude often prevails which is called the 'Florence Nightingale syndrome' in nursing. Traditionally, in the human services assertiveness is, wrongly, confused with selfishness. For that reason, learning to say 'no' is very difficult for many professionals, yet it is highly instrumental for controlling one's quantitative work-load and thus eventually for preventing burnout.

Corcoran and Bryce (1983) reported a significant decrease in levels of emotional exhaustion in social workers who received interpersonal skills training that focused on the affective component (i.e. reflective listening, personalisation, and empathy) compared with a non-treated control group. No significant effects were found in another treatment group that focused on the cognitive component of assertiveness (i.e. open ended questioning, reflection of feelings, paraphrasing, and summarisation). This result suggests that focusing on feelings is more effective than focusing on thoughts.

Promoting a realistic image of the job

Professionals usually enter their jobs with high hopes and expectations. These are reinforced by a set of publicly held beliefs and opinions about competence, autonomy, self-realisation, and so on, referred to by Cherniss (1980a, pp. 249–256) as 'professional mystique'. This ideal, which is expressed not only in the popular media but also in job advertisements, is actually highly unrealistic. As we have seen in previous chapters, high and wrong expectations fuelled by a faulty ideology are a major risk-factor for developing burnout. Hence, promoting a more realistic image of the job is a powerful preventive strategy, particularly to avoid an initial reality shock and thus reduce the risk of early career burnout (Cherniss, 1980a, p. 228).

It has been suggested that preparation should preferably take place in special training courses before students fully enter such fields as teaching (Cunnigham, 1983), mental health care (Pines and Maslach, 1978), or nursing (Kramer, 1974). A recent study showed that nurses who had received 'job expectancy training' were less burned-out than nurses who did not receive this training (Mickler and Rosen, 1994). In job expectancy training they learned that patients often resist being helped and they were taught what to do when that occurs. More specifically, the trained nurses expressed a greater sense of personal accomplishment and less depersonalisation than the untrained nurses. Quite interestingly, informal expectancy indoctrination by peers had a similar positive effect on nurses' level of accomplishment.

Balancing work and private life

Freudenberger called burnout 'the disease of the over-committed' or 'the super-achiever sickness', suggesting that burned-out employees are extremely dedicated to their work. For many of them there is no life outside work, which makes them extremely vulnerable to burnout. Or as Maslach (1982b, p. 104) puts it: 'When your whole world is your work and little else, then your whole world is likely to fall apart when problems arise on the job'. Thus, the lesson to be learned here is that a rich and varied private life is important since it complements the public life of work. It offsets the emotional strain of work and helps one to recharge one's batteries.

Dynerman and Hayes (1991) interviewed several hundred people and, based on these interviews, came up with practical suggestions on how to balance work and private life. For instance, setting up clear boundaries between job and home, physically as well as psychologically, limit job spillover, pursue leisure activities that are fun and rewarding, and spend more time in the company of others. Cherniss (1995) followed 26 young professionals who suffered from early career burnout over 12 years and found that those who recovered were more successful at balancing work, family and leisure:

> The most successful professionals considered family commitments and leisure pursuits to be at least as important as their careers. They didn't simply work to live, as the burned out professionals did, but neither they just live to work.   (p. 162).

Maslach (1982b, pp. 101–103) pays special attention to making the transition from work to home by introducing the notion of 'decompression'. Decompression, the gradual transition out of a high pressure underwater environment, is a technique used by scuba divers in order to avoid physical damage and unpleasant sensations caused by the release of nitrogen bubbles into the bloodstream. In a similar vein,

Maslach argues, people working in an environment of high emotional pressure need to 'decompress' before moving into the normal pressure of their private life. Decompression refers to any activity that occurs between working and non-working times and that allows one to unwind, relax, and leave the job behind before getting fully involved with one's private life. For instance, reading a book, gardening, window shopping, daydreaming, going for a walk, or taking a nap may all be used for this purpose.

Peer-support groups

Social support from colleagues is of great importance for the prevention of burnout and there are several ways to structure and organise support at work. Basically, regular staff meetings are important vehicles for giving and receiving social support. There may be a drawback, however. Pines and Maslach (1978) observed that staff meetings in a mental health institution were counterproductive because, instead of conferring openly with other staff about themselves, only patients were discussed in a negative, detached and intellectual way. This example shows that social support is not necessarily provided in regular meetings and that special occasions must sometimes be created: peer-support groups. But even such groups might be ineffective because of the contagious nature of burnout as is illustrated by Cherniss (1980a, p. 233) who warns: 'When professionals who are already burned out come together to discuss their feelings and experiences, the outcome could well be even more burnout, whether or not the gathering is euphemistically called a "support" group'.

We use peer-support group here as a generic term to denote any group of co-workers who come together on a more or less regular basis to exchange information, support each other emotionally, or to solve problems at work. These groups may vary greatly as to the extent to which they are formalised. At the one end of the spectrum one finds loosely organised groups of co-workers that discuss immediate concerns of the participants (e.g. Randolph, 1981), whereas at the opposite end one finds well-structured groups that use a specific, systematic approach (e.g. Balint-groups; see Rabinowitz et al., 1996). However, all peer-support groups have in common at least five elements. They provide:

- **Recognition** Sharing particular ideas, problems, or concerns with others is emotionally reassuring. Moreover, peers may serve as a basis for comparison since they provide a yardstick against which one can measure one's own feelings and actions.

- **Comfort** Peers provide a shoulder to cry on, or a sympathetic ear.

- **Help** Peers may offer direct aid or assistance (e.g. take over duties) or suggest practical solutions.

- **Insight** Support groups offer the opportunity to learn from each other. How did others handle that problem? What kind of resources did they use?

- **Companionship** Being together in an informal atmosphere and discussing common concerns strengthens the interpersonal ties. It is a pleasant escape from everyday routine and counteracts social isolation.

These five functions of peer-support groups roughly correspond with the types of social support that are generally distinguished: appraisal support, emotional support, instrumental support, informational support, and rewarding companionship, respectively. Because of the fact that peer-support groups cover practically all aspects of social support, they seem quite useful in preventing burnout.

The empirical evidence on their effectiveness is mixed, though. Brown (1984) showed that weekly held peer-support groups did not reduce burnout levels in nurses at the follow-up after 5 months. However, participants were more satisfied with their co-workers and supervisors. These findings are in agreement with Larson (1986), who evaluated a 12-week peer-support programme for hospice and oncology workers. On the other hand, Cooley and Yovanoff (1996) showed that a 1-month peer-support collaboration programme, albeit in combination with a stress-management workshop, was effective in reducing levels of emotional exhaustion and depersonalisation in special educators. The effects were sustained at 6 months and 1 year follow-ups. In a somewhat similar vein, Rabinowitz et al. (1996) observed that nurses who participated fortnightly in a Balint-group for 10 months displayed less emotional exhaustion and cognitive weariness at the end of the course.

In Balint-groups, peers discuss patients about whom they feel concerned, no matter why, and deliberate aspects of their work that they find personally troubling. The group, which is chaired by a leader, assesses the cases on their relevance to its members but does not examine their emotional states. Weiner et al. (1983) compared successful and non-successful peer-support groups of critical care nurses. They concluded that groups work best: if they are initiated in response to a need felt by the nurses; if the nurses have experienced the group leader as helpful in the past; if groups are highly structured; if early discharge of intense negative feelings is not allowed; if the group's problems are primarily interpersonal. Except for this final point, Balint groups seem to match pretty well with this description of successful peer-support groups.

Individual peer-support

What about employees who see their colleagues burning out? According to a survey of the American Psychological Association held in the early 80s, about 70% of psychologists knew colleagues who they believed were experiencing personal or emotional problems. Yet, only 36% had approached a colleague with their concerns (VandenBos and Duthie, 1986). This reluctance is probably caused by two interwoven processes that reinforce each other. Burnout candidates tend to deny their problems because they are ego-threatening (see Chapter 5), whereas their colleagues and supervisors tend to ignore and avoid their colleagues' problems because they fear embarrassment, hostility, and anger. Serious dilemmas might occur when, for instance, the quality of care of a colleague gravely deteriorates because (s)he is burning out. VandenBos and Duthie (1986) proposed a stepwise approach to dealing with a burned-out colleague at work:

- **Evaluate the problem** It takes less information to become concerned than it takes to be convinced and convincing. Hence, notes should be kept systematically about critical incidents, unprofessional behaviours, and so on.

- **Prepare before the meeting** Defensive denial is likely to be the initial emotional response. Instead of generally addressing the issue, specific behaviours that best illustrate one's concerns should be selected.

- **Listen, speak, and discuss** Key behaviours during the actual meeting with burned-out colleagues are attentive listening, responding in an empathetic way, and assisting in problem solving.

- **Follow-up** Meetings should be continued on a regular basis. The outcome of the process may be that the burned-out colleague looks for professional assistance or help.

Coaching and consultation

Coaching and consultation both refer to situations where expert help from a more experienced co-worker, manager or supervisor is offered to employees for work problems. Consultation pertains to a more or less unique event (e.g. how to deal with a particular client), whereas coaching refers to a series of such events (e.g. how to deal with aggressive clients). Unlike peer-support, an expert instead of one's peers plays the crucial role and unlike counselling, coaching and consultation focus on ordinary work problems, rather than on the personal problems of the employees. Generally, in coaching and consultation, interpersonal problems concerning recipients and/or co-workers are addressed. The primary aim of coaching and consultation is to support professional development. More specifically, it combines three functions:

- **Social support** The employee presents his or her work problem to the expert and jointly they seek to solve it. In so doing, not only is instrumental and informational support provided, but also emotional and appraisal support, and rewarding companionship.

- **Reflection** Coaching, in particular stimulates the employee to systematically and critically reflect upon his or her daily work in order to improve its quality.

- **Feedback** The expert provides detailed and personal feedback on the employee's performance. This not only improves motivation, but might also clarify one's role in the organisation, thus preventing or alleviating role problems.

The importance of coaching and consultation in the prevention of burnout is generally acknowledged because it not only offers social support but it also reduces qualitative workload and avoids potential role problems. Cherniss (1995) observed in a longitudinal study that guidance from colleagues was especially significant during the first year or two of the career. He concluded:

> The professionals who avoided early career burnout received helpful advice, information and feedback from older colleagues.   (p. 146)

The great need for coaching is illustrated by a survey among German human services professionals (Enzmann and Kleiber, 1989). It appeared that coaching was by far the most popular among various kinds of professional retraining, not only because it increased personal and professional development, but also because it was expected to reduce job stress.

Career planning

Most employees spend most of their working life in the same job. Thus, in order to keep the spark alive, career development is needed. Career development consists of two components: career planning and career management. The former is the responsibility of the individual, whereas the latter is the responsibility of the organisation (see later). Effective career planning includes two key elements:

- **Self-analysis** provides information about one's strengths, weaknesses, interests, and abilities and might include an assessment of one's current level of burnout.
- **Opportunity-analysis** identifies the range of organisational roles available.

Self-analysis and opportunity analysis constitute the basis for a strategic career plan that should be placed in the context of an overall balanced life. Because of radical organisational changes such as downsizing, delayering, mergers and acquisitions, the traditional view of a career as a series of upward moves is no longer realistic and should be abandoned in favour of a career based on a series of lateral moves. Unfortunately, for most professionals, career opportunities are limited because of the highly specialised nature of their jobs. For instance, it is quite difficult to imagine a career for a dentist in a private practice. It is therefore not surprising that many professionals feel 'locked' in their careers and are at risk of burning out. Although individual career planning, possibly supported by counsellor assistance or career planning courses, is certainly useful, it should be embedded in career management strategies at the organisational level (see later).

Specialised counselling

We now leave the field of prevention and enter the realm of treatment. As noted before, treatment includes employees who are actually burned-out: either they have serious difficulties in fulfilling their work roles, or they are on sick-leave. It is important to note that a gradual distinction exists between several individual intervention strategies, which is based on the seriousness and complexity of the employee's problems or symptoms. A preventive strategy such as, for instance, career planning may also be used for treatment purposes among employees who experience serious problems at work. Specialised counselling is performed by professionals such as general practitioners, social workers, counsellors, or occupational physicians for employees who are in a temporary crisis and for those who have somewhat more serious problems. Finally, psychotherapeutic treatment of burnout is conducted by highly specialised professionals – usually psychiatrists, psychotherapists, or clinical psychologists – who deal with the most complex and severe cases.

Quite remarkably, the literature on counselling of burnout is rather meagre. Typically, most authors do not go beyond recommending rather non-specific strategies or techniques such as didactic stress management, relaxation, or RET (Edelwich and Brodsky, 1980; Kaslow, 1982). An exception has to be made for Van der Klink and Terluin (1996), who proposed a specific approach for counselling burned-out employees. They consider burnout as a temporary crisis in coping that leaves the employee bewildered and feeling out of control ('I do not understand how this could happen to me'). The usual advice to take some rest and to retreat from one's duties is false, they argue. It reinforces a passive attitude because it suggests that healing just 'happens' and that it is thus beyond one's personal control. Instead, Van der Klink and Terluin (1996) advocate an active approach that restores the employee's control over his or her (working) life and stimulates employees to take responsibility for their own behaviour. In other words, rather than passively waiting for it, healing is a 'job' that has to be actively accomplished. More specifically the authors distinguish three phases in the counselling process:

- **Managing the crisis** The client should be made to understand the problem, to accept it, and to consider it as a problem that can be solved by his or her own effort. In order to make the client understand, a rationale is presented, for instance, in the form of a balance that has been disturbed: one's efforts have temporarily outweighed one's resources. Furthermore, a perspective is offered, which should strengthen the client's feeling of control: for example, that 75% of burned-out employees recover within 12 weeks. Next, an activation programme is carefully drawn up that consists of alternating periods of rest and activity. Finally, the client's burnout is positively labelled as a warning sign.

- **Solving the problem** The client makes an inventory of (work) problems by using self-monitoring techniques. It is important to describe these problems as specifically as possible, preferably in behavioural terms. They are prioritised, further analysed, and specific techniques are used to solve them (e.g. time-management, relaxation). Simultaneously, the client's stress-resilience is increased by a physical exercise programme.

- **Returning to ordinary life** The client is prepared to take up his or her usual duties again by gradual and systematic exposure. This process is carefully monitored by the counsellor and guided by two scenarios that the client has been asked to prepare: an optimistic and a pessimistic scenario. This technique permits the therapist to stimulate overly cautious clients by emphasising the optimistic scenario, and to slow down those who tend to rush too quickly by emphasising the pessimistic scenario. In order to prevent the client from burning out again, relapse prevention is recommended: while anticipating future problems the client rehearses adequate coping behaviours.

Psychotherapy

Despite the claim of Sigmund Freud almost one century ago that work and love ('Arbeit und Liebe') are man's strongest ties to reality, psychotherapeutic treatment has almost exclusively been focused on the latter at the expense of the former. Only very few examples exist of psychotherapeutic treatment of clients who suffer from job-related stress. One notable exception is the Sheffield Psychotherapy Project that was successful in treating white-collar, professional and managerial employees who suffered from depressive or anxiety disorders with prescriptive (cognitive/behavioural) and exploratory (relationship-oriented) therapy (Shapiro and Firth, 1987; Shapiro et al., 1990). Unfortunately, this project did not deal explicitly with burnout.

To date, Lowman (1993) presents the most comprehensive overview of psychotherapeutic treatment of 'work dysfunctions' which he defined as '. . . psychological conditions in which there is a significant impairment in the capacity to work caused by either characteristics of the person or by an interaction between personal characteristics and working conditions' (p. 4). Burnout, according to Lowman (1993, pp. 107–142), is a work dysfunction which is characterised by a particular pattern of overcommitment that refers to too intense an identification with and involvement in the work role. He posits that for treating burnout the usual psychotherapeutic (cognitive-behavioural) techniques can be applied. However, they have to be framed within the context of the client's work problems. Lowman (1993) presents the following guidelines for treating burned-out clients:

- **Use humour and mild sarcasm** This helps to put the client's controlling, compulsive, and grim behaviour into perspective. Paradoxical interventions may have a similar purpose (e.g. a 'duty' to play).

- **Give the client the permission to enjoy himself or herself** Usually, burned-out clients feel that they first have to 'perform' before they allow themselves to relax or enjoy something.

- **Emphasise that the client has a choice** For many burned-out clients the idea of choice (career, recreational activities) does not occur naturally.

- **Use homework assignments** Since many burned-out clients are quite conscientious and duty-bound, homework assignments are deemed to be successful ('For every 5 hours of work, you must spend at least 30 minutes waking time in recreational activities').

- **Anticipate that treatment easily becomes 'work'** As a consequence (s)he may become upset about the lack of progress ('I understand this, but what am I going to do to change it?').

- **Do not communicate rejection** Rejection is generally well known, particularly in close interpersonal situations. It reinforces what overcommitted persons are skilled at doing without realising: alienating others.

- **Do not push too hard for change** Pushing hard might lead the client to abruptly end the therapy, particularly as his or her vulnerability becomes exposed.

Lowman (1993, p. 142) also warns against symptom contagion:

> Because many therapists are themselves overworked, perfectionistic, and overcommitted individuals who get too little pleasure out of life, work with clients who have similar problems may arouse some of the therapist's own anxieties about the work role.

Therefore, providing a role model of an unpressured, productive worker whose work exhibits a positive drive may be the first, and probably most difficult, challenge for the therapist.

In contrast to Lowman (1993), who does not present a specific psychotherapeutic approach to burnout, there is some experience with treatment of burnout victims in The Netherlands (Schaap *et al.*, 1995; Hoogduin *et al.*, 1996a). The Dutch treatment programme is based on the principles of cognitive-behavioural therapy and relies heavily on self-control procedures, including self-monitoring. Regaining self-control is considered essential since most burnout victims feel bewildered, confused, powerless, and out of control. The programme is illustrated in Box 6.5 which summarises the treatment of Mr Dijkstra, the disappointed teacher whom we met previously (see Box 2.1). It consists of the following four phases:

- **Reduction of target complaints** The five most often observed target complaints are: mental and/or physical exhaustion; worrying and the inability to relax; sleep disturbances; irritability; physical distress symptoms. Not surprisingly, these complaints correspond with the symptoms of work-related neurasthenia that were previously discussed in Chapter 3. These target complaints are treated with well-known cognitive-behavioural techniques such as gradual activation (exhaustion), worry sessions (worrying), cue-conditioning (inability to relax), relaxation (sleep disturbances), and stimulus control and response cost (irritability).

- **Understanding one's personality** It appears from clinical experience that two different personality types seem to run a special risk of burning out: individuals with a narcissistic personality and those with a sensitive-compulsive personality. These personality types correspond with the prototypes of the assertive, dominating, and outgoing employee and with the conscientious, diligent, and accurate employee, respectively. Using techniques such as cognitive appraisal or RET, clients learn to deal better with their problematic personality traits that have caused trouble in the past. These cognitive techniques are supplemented with behavioural exercises in which, instead of the usual negative thoughts, alternative positive thoughts are produced.

- **Tackling work problems (by skills training)** If the client is on sick-leave, a rehabilitation plan is drafted in cooperation with the occupational physician and the personnel officer, as soon as the target complaints are significantly reduced. That plan maps out in what gradual way, at what time, and with what duties the client returns to work. The client's behaviour at work is carefully monitored *in vivo* by using a diary, which may reveal role problems, work overload, interpersonal conflicts, or lack of social support. The client is encouraged to tackle these problems actively, for instance by talking directly to his or her supervisor. Usually, there are serious and recurring problems in time-management or other social skills such as assertiveness, chairing meetings, or speaking in public. Accordingly, in this stage of the treatment programme, skills are rehearsed that are useful in reducing stress at work.

■ **Anticipating the future** In the final stage, relapse prevention is applied. That is, the client learns how to identify particular situations that might trigger undesired responses and learns how to cope with them. In addition, a concrete plan for a more healthy lifestyle is drafted. The treatment programme is concluded by discussing from an existential perspective the role that work plays in the client's life. Most burnout victims have put all their energy into their jobs at the expense of their private lives. Therefore they should find a new balance between work and non-work, and between their public and their private lives.

The above-mentioned psychotherapeutic treatment programme is unique not so much because it includes new techniques to combat burnout, but because a set of specific, commonly used techniques is systematically applied to clients who have similar target complaints, personalities and work problems. Essentially, the programme moves from a more specific level (i.e. reducing particular target complaints) via personality and work problems, to a more general level (i.e. how to rearrange one's life in order to find a new balance). The programme is comprehensive not only because it includes various techniques and levels, but most of all because it also combines healing and rehabilitation.

Referral

Strictly speaking, referral is not a treatment strategy but the possible outcome of a systematic evaluation of the client's problems. It is particularly relevant for personnel officers, social workers, occupational physicians, and occupational nurses who are confronted with burned-out employees. Effective referral not only requires careful assessment of the employee's problems, it also requires detailed and up-to-date knowledge about local programmes and initiatives that are effective in reducing burnout. In addition, formal rules and regulations should be considered, including workers' compensation and insurance claims. Lowman (1993, p. 49) presents a decision tree that may be used as a tool to adequately refer the client, for instance, to a career counselling service, a burnout workshop, specialised counselling, or individual psychotherapy.

Guidance and assistance with rehabilitation

Ideally, rehabilitation, the planned return to the previous job after burnout, should be an integral part of every treatment programme, as was the case with the counselling and psychotherapeutic programmes we discussed above. Rehabilitation is crucial since many burned-out employees either do not go back to their previous jobs or do not manage to stay there. In The Netherlands, only 55% of the burned-out

Box 6.5   Psychotherapeutic treatment of the disappointed teacher

In the *first session*, it was explained to Mr Dijkstra how burnout develops, what its symptoms are and what the prognosis is like. Mr Dijkstra's symptoms were interpreted as distress symptoms that indicated that he had exhausted his energy resources. It was explained how to monitor his level of anxiety by using a stress-diary: every hour he had to assess on a special form the level of anxiety that he was currently experiencing on a ten-point scale running from deep relaxation (1) to panic (10). He was also to write down what caused that anxiety and what strategy he had used to cope with it.

In the *second session* the stress-diary was analysed and it appeared that anxiety started with fatigue and headaches. In addition, Mr Dijkstra worried about school and about himself. He was instructed how to conduct a daily half hour private 'worry session' that was to take place in a particular room at a particular time. The topics Mr Dijkstra worried about were written down and solutions were suggested. In order to stop worrying outside this half hour period, various activities were suggested that were incompatible with worrying such as talking, motor activity, and paying attention to his spouse.

In the *third session*, progressive relaxation was rehearsed and audio-taped. Mr Dijkstra was to practise twice a day. Over the previous week he had learned through self-monitoring that anxiety also starts with feeling restless and pressure in the neck. It was agreed that as soon as these symptoms occurred Mr Dijkstra should start tinkering at his garage, going for a walk, or listening to music.

In the *fourth session* the relaxation exercises were evaluated positively since they reduced Mr Dijkstra's level of anxiety. During this session cue-conditioning was rehearsed: relaxation was linked to the clenching of Mr Dijkstra's right fist, so that, in difficult situations, he could relax quickly simply by clenching that fist. In addition, a self-control programme was designed for his irritability. It was agreed that Mr Dijkstra's wife should register how often he was unkind to his children. Such irritable behaviour was penalised by carrying out some of his wife's bothersome or dull duties (response cost).

At the beginning of the *fifth session*, Mr Dijkstra reported that the idea of returning to work that had been discussed in the previous session had caused a lot of anxiety. However, in the past week he had visited his school and talked to his colleagues. It was less stressful than he had anticipated although answering their questions about his present health status was rather difficult. He also arranged a meeting with the principal to discuss his future duties. It was agreed, in accordance with the occupational physician, that Mr Dijkstra would return to work. During a period of 2 or 3 months, the number of working hours would be gradually increased to 20 hours. The RET philosophy was explained in order to tackle Mr Dijkstra's irrational ideas and his resulting depressed mood.

In the *sixth session* Mr Dijkstra reported a slight relapse. The principal had asked him for assistance on a particular issue, but it appeared to be rather stressful for him. He regretted that he had agreed but he did not tell the principal. This particular situation was analysed using the RET-scheme (see Box 6.3). Three irrational beliefs emerged: 'I have to do everything perfectly'; 'I have to do this kind of work, since nobody can do that better'; 'Everybody must like me'. These beliefs were challenged and alternatives were formulated. Mr Dijkstra was encouraged to phone the principal and to tell him that he wasn't ready to do that particular job yet.

In the *seventh session* – the sessions were now held fortnightly – the telephone call to the principal was discussed. It went all right, the principal had been quite understanding and cooperative. Meanwhile, Mr Dijkstra had not yet returned to school. The necessary requirements for doing so were reviewed: Mr Dijkstra should be able to identify anxiety, relax in time, not to take on too much work simultaneously, say 'no' at the proper time, and he should not feel responsible for everything.

In the *eighth session* Mr Dijkstra stated that he still did not feel ready to return to work. His stress-diary was analysed and RET continued.

In the *ninth session*, Mr Dijkstra claimed that he was now able to relax. He also felt less fatigued. He had observed that he slept poorly when something special was about to happen the next day, for instance, a party or a family visit. Mr Dijkstra was recommended to use RET in order to analyse what thoughts were causing that anxiety. Furthermore, he was encouraged to look for stressful situations in order to practise cue-conditioning in real life settings.

In the *tenth session* situations were discussed in which Mr Dijkstra should say 'no'. Also situations were analysed where he should be more clear in saying what he really wanted and how he actually felt (displaying assertiveness).

In the *eleventh session* Mr Dijkstra announced that he wanted to go back to school for a couple of hours each morning. Once more, the necessary requirements for returning to his previous job were reviewed and rehearsed.

In the *twelfth session* he reported another relapse because the principal had told him that his absence had caused a lot of trouble because others had had difficulties taking over his duties. Further analysis showed that again, Mr Dijkstra's frustration went back to his perfectionism. Also he was annoyed by his colleagues who didn't really care.

The *thirteenth session* was spent entirely on time-management. Mr Dijsktra showed his daily planning, he had registered how much time he spent on each duty and prioritised his activities. The most important duties were done first. Furthermore, he had planned to delegate particular tasks to others. Meanwhile, Mr Dijkstra was working half-time. He agreed with the principal that he would teach more and do less administrative work.

The *fourteenth session* was used mainly for relapse prevention. Mr Dijkstra learned to use signals that indicated that he was asking too much of himself in such a way as to reduce his work pace or his work load. By now he was able to effectively cope with anxiety by using cue-conditioning. He also successfully performed relaxation exercises during the breaks at school and used rational self-analyses nearly every day.

At the *follow up* after three months Mr Dijkstra was still doing well. His scores on the Symptom Checklist 90 as well as on the MBI had returned to normal levels again.

Source: Schaap *et al.* (1995)

employees who have been on sick-leave return to their previous jobs; 1 year after reporting sick only 40% are still working in that same job (Schroër, 1993). Furthermore, it was observed that active rehabilitation practically does not exist: most burned-out employees were treated by their family doctor and mental health professionals were involved only in about one quarter of all cases. Physicians usually dealt with burnout by recommending the individual to take a rest, prescribing tranquillisers, or inviting the patient to come back for a talk. Only in one out of every five cases is a contact established between the organisation and the health care professional who is responsible for treatment. Quite remarkably, the initiative for that contact is generally taken not by the treatment professional but by the organisation.

The following conditions and recommendations are important for properly guiding and assisting burned-out employees to return to their jobs (cf. Hoogduin *et al.*, 1996b):

■ **Symptom reduction** Of course, burnout levels should be markedly reduced.

■ **Self-confidence** Relapse prevention may help to restore and increase self-confidence.

■ **Rehabilitation plan** All professionals involved (e.g. personnel officer, supervisor, general practitioner, occupational physician, counsellor) should agree upon a plan of how to reintegrate the employee in his or her previous job. The plan may include such issues as gradual exposure to job demands, limited number of initial work hours, and adaptation of duties.

■ **Self-monitoring** The employee should systematically monitor his or her symptoms during the rehabilitation process.

■ **Employee-monitoring** The rehabilitation process should also be monitored by the health care professional responsible for the employee. Information should be collected from other relevant persons such as supervisors and personnel officers.

■ **Time-perspective** A short time-perspective, adapting to the previous work role, should be distinguished from a long time-perspective – e.g. a career in a different field.

Changing jobs

Rehabilitation might take the form of changing jobs, that is, integration in a fresh work role. In spite of all efforts, the employee may come to realise that changing jobs, or maybe even leaving the field, is the only option available in order to sustain health. This is a very serious decision that should be based on careful consideration of the underlying reasons and the alternatives at hand. Maslach (1982b, p. 107) warns:

> If the change is more superficial than real, then the risk of burnout is not really reduced. Going into the same type of job and handling it in the same way as before does not represent progress.

A well-planned change of jobs may well be a positive step in personal growth, protecting the employee from future burnout. Cherniss (1995, pp. 181–190) argued that significance is the key factor of such a new job. He found that of the five professionals who recovered from early career burnout, four changed their jobs within the first 3 years. This change brought about a large improvement in how meaningful their work was in terms of impact, intellectual challenge, and professional growth.

6.4 INTERVENTIONS PRIMARILY AIMED AT THE ORGANISATION

Organisation-based interventions for reducing stress and burnout may focus on: surveillance (i.e. conducting a stress audit and a psychosocial check-up); removal or reduction of stressors (i.e. improving the job content and work environment, better time scheduling, improving conflict management, communication, decision-making, and organisational development); improving the fit between employee and organisation (i.e. career management, retraining, anticipatory socialisation, management development, and outplacement); institutionalisation of projects and services (i.e. corporate fitness programmes, enrichment of occupational health and safety services, and employee assistance programmes).

Stress audit

Employee surveys or stress audits may be used to take the 'stress-temperature' of the organisation by comparing employees' scores across units, locations, occupations, jobs, and so on. Starting in the early 1950s American companies such as Sears, IBM, and AT&T developed a survey tradition using 'climate surveys', 'attitude surveys', 'opinion surveys', or 'employee reaction surveys'. Such surveys are designed to elicit employee reactions and preferences in order to assist management in developing action strategies that might improve organisational effectiveness and employee well-being. In addition, by carrying out surveys, the organisation recognises that job stress and burnout are legitimate problems.

Basically, instruments similar to those that are used for personal screening can be included in a stress audit. However, the focus is now on organisational aspects such as risk-factors in the work-environment rather than on personal aspects such as the employee's way of coping with stress. For a recent example of a stress audit in a large telephone company see Judge (1994). When carrying out a stress audit some issues have to be taken into account:

- **Anonymity** The survey should be completed anonymously in work time.

- **Periodicity** Surveys should be carried out on a regular basis so that comparisons across time can be made.

- **Specificity** The survey should be tailor-made and as specific as possible to ensure employee compliance.

- **Norms** should be specific for the particular job, organisation, or profession involved.

- **Feedback** Results must be fed back to the participants and openly discussed with them.

- **Commitment** Management should be committed to the survey process, including the follow-up with specific actions.

Pines and Maslach (1980) used a stress audit in a day-care centre and found that burnout was related to the unstructured and non-directive quality of the care programme. After a series of meetings to discuss the implications or the results of the audit, the structure of the programme was changed towards less permissiveness. At the 6-month follow-up, this strategy proved to be effective in reducing burnout levels.

Psychosocial check-up

Analogously to a periodical physical check-up, some authors have proposed that organisations should institutionalise a voluntary 'burnout check-up', for instance, every 6 months or so (Cherniss, 1980a, pp. 230–231; Maslach 1982b, p. 123). If a high level of burnout is observed, the particular employee may be referred for further psychodiagnostic assessment, or he or she may be recommended to attend an anti-burnout workshop or to seek treatment.

Improving the job content and the work environment

Measures to improve the job content and the work environment are basically directed towards reducing quantitative and/or qualitative work overload. The most straight-forward way to reduce quantitative overload is to hire more employees, for instance, in order to improve the organisation's client-to-staff ratio (Pines and Maslach, 1978). This option is costly and therefore unlikely to be implemented, unless a cost-benefit analysis reveals that the costs associated with burnout – turnover, sick-leave, mental health claims – exceed the expenses for hiring more employees. For two reasons, Cherniss (1980a, p. 159) is rather critical about hiring more employees. First, practical experience suggests that only drastic measures are effective. For instance, a positive effect on teachers' workload can only be expected when the number of pupils in the classroom drops by at least a dozen or so. Secondly, hiring additional employees is not an effective strategy as long as there are other problems with the content of the job or with the psychosocial work environment.

Broadly speaking, three alternative types of strategies can be used to reduce workload and thus counteract burnout: job redesign; clarifying the employee's role; improving the physical work environment. Box 6.6 presents some examples of **job redesign** to reduce burnout.

In many organisations, formal job descriptions outlining the employee's duties are not available. This increases the likelihood of role problems which, in their turn, might foster burnout (see Chapter 4). Hence, detailed job descriptions, which should preferably be behaviour-based, are expected to be effective, especially in alleviating role ambiguity (Maslach and Jackson, 1984a). Ivancevich and Matteson (1980, pp. 210–211) advocate **role clarification** as a technique to analyse discrepancies in mutual role expectations of supervisors and subordinates. This is achieved by systematically answering such questions as: 'What do you think is expected from you?' and 'What do you expect from your supervisor?' Role clarification is fostered by regular performance feedback and performance evaluation.

Although the physical work environment is not the main cause of burnout, it might aggravate particular distress symptoms, such as headaches or irritability. Therefore, attention should be paid to the **improvement of the physical work environment**, for instance through focus groups. In such groups the expertise of employees who work in the same job is used to generate ideas and solutions about how to improve working conditions that are causing stress (cf. Ross and Altmaier, 1994, p. 117).

Time scheduling

Since protracted and intensive contact with recipients is considered to be a root cause of burnout, many suggestions have been made for organisations to change their employees' time schedules. Most proposals aim to either reduce the number of working hours, or reduce the time spent face-to-face with recipients (e.g. Cherniss, 1980a, pp. 238–239; Maslach, 1982b, pp. 125–126; Pines and Maslach, 1978; Cherniss and Danzig, 1986):

- **Mental health days** Organisations can provide such days, which differ from sick-leave, and on which employees are given some time off to unwind and recharge their batteries.

- **Sabbatical leave** This is an extended period of time, usually a month to a year, when the provider is freed from his or her regular work to do something else that is job enriching. In some countries sabbatical leave is particularly common in education.

Box 6.6 Job redesign to reduce burnout

Job enlargement: adding duties or responsibilities to the current job. As in many other fields, a rather rigid division of labour exists in the human services. For instance, it is not unusual in a social work agency that some professionals perform the intake of new clients, whereas others treat them, and still others manage contacts with outside agencies, or are involved in prevention projects. Maslach (1982b) proposed to enlarge jobs, for instance, by combining intake and treatment, or prevention and liaison.

Job enrichment: restructuring a job so that it is more meaningful, challenging, and intrinsically rewarding. For instance, in a Swedish psychogeriatric clinic individually planned nursing care was introduced on a number of wards. That is, each patient was assigned a particular nurse who was responsible for all nursing tasks. Traditionally, patient-care was delivered by many nurses who worked in various shifts, with no one having special responsibilities for the continuity of care. At the 1 year follow-up, levels of burnout of nurses in the experimental wards had dropped significantly, compared with the traditional control wards (Berg et al., 1994). However, two recent Dutch studies on job redesign in nursing failed to confirm this positive result. One study compared psychiatric nurses who worked according to similar nursing principles as in the Swedish study with a control group that kept working in the traditional way (Melchior et al., 1996). In the other Dutch study the effects of differentiated practice and specialisation in community nursing were investigated (Jansen, 1996). Based on an assessment of *complexity* of nursing care, patients were assigned either to a community nurse (complex cases) or a community nurse aid (less complex cases). This was called differentiated practice. Based on an assessment of the *nature* of nursing care, patients were assigned to a nurse or nurse-aid depending on his or her area of expertise (e.g. cardiac diseases, cancer, neurological disorders). This was called specialisation. Compared with the control group, both experimental groups did not show a decrease in burnout. In fact, burnout levels increased during the 1 year study period in all groups. The authors speculated that the workload had increased because management had put more emphasis on productivity and efficiency. Besides, information leakage had occurred through nurses who switched from the experimental to the control wards and vice versa. These problems demonstrate once more how difficult it is to perform a scientifically sound study in a practical setting.

Job rotation: periodically changing jobs or duties. For instance, in order to share the patient load, psychiatric nurses may change periodically from 'difficult' to 'easy' wards, and vice versa (Pines and Maslach, 1978). In a similar vein, Cherniss and Danzig (1986) proposed to distribute the 'dirty work' evenly, that is, to spread out the most unpleasant duties through rotation of assignment. 'Dirty work' may pertain to particularly difficult and unrewarding clients or programmes in the human services, or to certain courses or committees in teaching.

- **Retreats** A system of retreats and workshops outside the institution for experienced workers may counteract the burnout process.

- **Encourage part-time employment** Working in a part-time job makes it easier to balance work and private life (see above).

- **Discourage excessive overwork** Supervisors should monitor the working hours of their subordinates and actively discourage 'heroic' behaviour such as working weeks of 50 or 60 hours, or more.

Management Development

Managers have two key roles aimed at the development of healthy work environments (Quick *et al.*, 1996):

- **Referent leader** They serve as a role model by practising the principles of an occupationally healthy lifestyle, which requires skills in stress-management, self-awareness, communication and conflict management, time-management, and so on.

- **Triage agent** That is, managers must determine when someone needs help and then refer the employee to the proper service. Of course, the manager should have resources to which people in need may be transferred (e.g. Occupational Health and Safety Service).

In order to fulfil these roles, managers should display particular characteristics such as openness, systematic thinking, creativity, self-efficacy, and empathy. Managers who fit this profile have been called 'healing managers' (Lundin and Lunding, 1994). However, according to Cherniss (1980a, pp. 240–243), instead of reducing, many managers and supervisors foster burnout because: they do not know about the psychological consequences of many of their decisions; they lack general management skills (e.g. prioritising, delegating) and specific interpersonal skills (e.g. active listening, expressing empathic concern); and last but not least, they work under considerable pressure and are therefore in danger of burning out themselves!

Management Development (MD) programmes may solve at least some of these problems. The primary method for implementing MD is through management education and management training. In addition, Cherniss (1980a) proposes to provide managers and supervisors with feedback about their leadership behaviour by regularly surveying their subordinates, for instance, as part of a stress audit. He claimed that such survey feedback has been successful in schools where it appeared that social leadership was more appreciated and effective than instrumental leadership in preventing job stress and burnout.

Career management

As previously noted, career management refers to the organisation's responsibility for developing the employees' careers. The basic aim of career management is to keep employees fit and productive by providing them with new challenges, thereby preventing occupational 'locking-in' and its possible consequence – burnout. Career management consists of an institutionalised set of rules and procedures that cover such areas as recruitment, selection, placement, development, and promotion. Organisations differ considerably as to the extent to which these rules and procedures are formalised. In human services organisations professional human resource management systems have been introduced only relatively recently.

Some organisations have career **management systems** that include training programmes to help employees understand the value of lateral moves, identify skills that transfer between job families, and take initiatives to make career moves. Other organisations maintain career resource centres that provide information on educational opportunities and self-study instruments, and include a computerised information bank on career planning. More specifically, Cherniss (1980a, p. 244) proposed that organisations should institutionalise a **mentor system** for newcomers that would act as a double-edged sword in preventing burnout. On the one hand, newcomers would be supported by a more experienced colleague in finding their way around in their new work environment and in solving problems at work, including those of a more emotional nature. On the other hand, being a mentor for a senior employee constitutes an alternative challenge that is potentially rewarding. Thus, the introduction of a mentor system might prevent early career burnout as well as late career burnout.

Retraining

Essentially, retraining improves the employee's professional qualifications so that qualitative work overload is reduced and burnout is less likely to occur. Social and interpersonal skills, but not technical skills, are often lacking, particularly among human services professionals in the early phases of their careers. Therefore, in-service training and training-on-the-job are recommended to counteract burnout (Cherniss, 1980a, pp. 215–220; Maslach, 1982b, pp. 136–141). It has been suggested that former recipients should be included in these training courses for role play and group discussion. After all, they are the experts *par excellence* who can provide inexperienced workers with valuable information and feedback.

More experienced employees may find it also increasingly difficult to deal with some aspects of their jobs because of rapid changes. For instance, Enzmann and Kleiber (1989) argued that in Germany in the late 1980s human services professionals were confronted with novel types of recipients (e.g. poor illiterate immigrants, sexually abused children) and new kinds of problems (e.g. unemployment, racism). Accordingly, there was a great need to retrain, not only in the strict professional domain. The authors observed that retraining courses in collaborative practice, communication, and conflict management were among the most popular. It appeared that professionals with the highest burnout levels attended these courses and that these levels dropped significantly to about the average level following the retraining programme.

Corporate fitness and wellness programmes

The reasons for institutionalising fitness and wellness programmes may be manifold: decreasing health care costs, improving employee health status, increasing productivity, and improving labour-management relations. More specifically, corporate fitness and wellness programmes may focus on one or more of the following targets (Schreurs *et al.*, 1996): control of high blood pressure, cessation of smoking, weight reduction, physical fitness, reduction of lower back pain, health and safety education, reduction of alcohol use, and stress management.

Stress management does not seem to be the most popular purpose since it is included in only about one quarter of the American programmes (Fielding, 1989). Instead, most programmes focus on reducing cardiovascular risk-factors such as blood pressure, smoking and obesity. In a recent review of the financial aspects of employee fitness programmes Shephard (1996, p. 48) concluded that these programmes 'appear to yield financial benefits that more than match the corporate costs of such initiatives'.

Although a number of programmes have been described in some detail (e.g. Batman, 1994; Maes *et al.*, 1992), effects on job stress and burnout were not systematically considered. In their critical review Kasl and Sexner (1992) concluded that the effective programme components have not yet been identified. It is nevertheless assumed that participating in a comprehensive corporate fitness and wellness programme might reduce employee burnout because it improves physical fitness and decreases vulnerability to job stress.

Anticipatory socialisation

Promoting a more realistic image of the job is a way of counteracting burnout through anticipatory socialisation (see above). Another way is to institutionalise a **realistic job preview**, a recruitment procedure that involves exposing applicants to the reality of the workplace before they are eventually hired. It is estimated that such previews may reduce employee turnover by about one third (Ross and Altmaier, 1994, p. 108).

Gradual exposure of employees to the demands of the job was recommended by Cherniss (1980b, p. 161), who described a socialisation approach that is used in public health nursing. When fresh nurses are hired, they are not assigned to patients. Instead, they attend seminars and workshops and accompany experienced nurses on their rounds. Gradually, the novices begin to assume more responsibility. A similar socialisation programme that was developed for preventing burnout was carried out amongst Canadian social workers entering child welfare (Burke and Richardsen, 1993). Workers were hired in batches so that small groups of about five individuals could be created. During the first 6 months of employment the worker's caseload was gradually increased to 60% of the usual caseload. In addition, a training programme of 1–2 days every 2 weeks was followed, and active social support for the groups was provided by their supervisors. Qualitative data suggests general satisfaction with the programme. For instance, supervisors felt that the level of skill during the first 6 months was comparable to that achieved in 1 year under the traditional orientation programme.

Conflict management, communication, and decision-making

Much frustration, stress and burnout is a response to bad conflict management, poor communication, or inappropriate decision making. Because of the complex social and professional nature of most jobs conflicts may easily arise. Quite often, roles are not clearly defined, job descriptions are lacking, expectations of supervisors are ambiguous or conflicting, and professional domains and competencies are ill-defined. In addition, the usual bureaucratic hassles and organisational policies cause conflict, political in-fighting, professional rivalry, jealous battles over turf, and red tape. During the past decade, conflicts have sharpened because of budget cuts, layoffs, mergers, and acquisitions. Moreover, often conflicts are either avoided because institutionalised conflict resolution mechanisms are lacking or they are personalised because employees are not considered as occupants of particular roles.

As far as a proper flow of bottom-up communication is concerned, a periodically carried out stress audit may be useful for providing information about particular psychosocial risk factors for burnout. In addition, formal top-down communication through periodically issued bulletins, intranet, or plenary meetings is increasingly important in today's large-scale, complex, and bureaucratic organisations.

Ideally, such communication channels should be embedded in a system of participative decision making. Jackson (1983) showed that those staff members of hospital wards who had an opportunity to offer input in regular staff meetings had an increased sense of influence and experienced less role stress and emotional strain than those staff members who could not participate in decision making because no such staff meetings were held. However, it is crucial that management is committed to participative decision making and the decisions should involve serious matters instead of being frivolous decisions, for example, about the colour of the waste paper baskets or where to put the new drinks machine.

Cherniss (1995, pp. 156–159) not only proposed to train staff in interpersonal problem solving skills, but also to pursue a more comprehensive strategy: the improvement of their 'organisational negotiation skill'. That is, professionals should learn effective ways to approach organisational conflicts and hassles and to view such systems problems in a more sophisticated, less personalising manner. More specifically, Cherniss (1995) distinguishes three components of organisational negotiation skill: the ability to avoid and resolve conflicts with co-workers and supervisors; the ability to generate organisational support for one's initiatives; a problem-solving attitude toward organisational difficulties (i.e. analytical detachment and thoughtful reflection).

Organisational Development

Organisational Development (OD) is a programme of planned interventions that should improve the internal operations of an organisation. Usually, its main focus is on improving quality, cost-effectiveness, or increasing efficiency: at best, stress reduction is an intended by-product. OD is both a methodology and a loose guidance system for helping organisations to make healthy changes. As a methodology it follows a step-wise approach (see below), as a guidance system it includes various techniques such as survey feedback, training, and team development.

Golembiewski and Rountree (1991) described an OD programme that wa carried out in a chain of nursing homes to improve service and profits, and to reduce burnout by releasing human potential for collaboration. In five experimental homes, pairs of directors of nursing and the respective local chief executive officers were formed. These pairs received a $2\frac{1}{2}$ day course of intensive training in team-building that included, amongst other things, a written action plan on how to improve their teams. After 1 year, the levels of burnout in the experimental pairs had dropped significantly, compared with five matched pairs from other sites who didn't part-icipate in the training. Unfortunately, levels of staff-burnout of the nursing homes involved were not reported.

Van Gorp and Schaufeli (1996) carried out an OD programme to reduce burnout in four large Dutch Community Mental Health Centres, using a four-step participatory action research approach (see Box 6.7).

Box 6.7   An OD programme to reduce burnout in Community Mental Health Centres

---

Preparation   Commitment of management was ascertained, a task-force was established in each Centre, and staff were informed at a plenary 'kick-off' meeting.

Data collection, diagnosis and planning   Surveys and interviews were carried out and analysed, and absenteeism records were studied. In addition to specific problems in each of the three Centres, three common problems emerged: work overload, poor team leadership, and tensions between professional staff (psycho-therapists, social workers, psychologists) and supporting staff (secretaries, desk clerks).

Action   Due to local differences in problems to be tackled and in the local dynamics of change, somewhat different actions were taken in each Centre. For example, the introduction of a system for standardising and allocating case-load, opening of communication channels between top management and shop floor, improving collaboration between professional and support staff by introducing a computerised data-information system, clarifying roles and duties, and improv-ing housing conditions.

Evaluation   After $1^1/2$ years a follow-up was conducted. Although the employees were satisfied with the way most problems were tackled, levels of burnout had not decreased markedly. However, in one Centre registered absenteeism had dropped significantly, whereas in the two remaining Centres psychosomatic com-plaints decreased substantively.

Source: Van Gorp and Schaufeli (1996)

Gold and Roth (1993, pp. 185–193) propose a similar OD programme for schools to reduce teacher burnout that includes three phases: organisation and planning; implementation; maintenance and evaluation.

Institutionalisation of Occupational Health and Safety Services

In recent decades, the importance of Occupational Health and Safety Services (OHSSs) has grown considerably. In Europe, the institutionalisation of OHSSs was facilitated by the introduction of new occupational health and safety legislation (de Gier, 1995). Despite national differences, labour policy was reformed along similar lines in Europe. First, new laws and regulations have broadened the scope of traditional legislation by including employee well-being in addition to occupational health and safety. Second, an integrated approach of these three areas is advocated. Third, health, safety and well-being at work are now regarded as the joint responsibility of employer and employee. So, instead of a passive victim who needs protection, the employee is considered an active agent who has a personal responsibility to protect him- or herself. Finally, a proactive and preventive strategy is emphasised rather than the usual reactive and curative strategy. As a consequence, the role of OHSSs has changed accordingly.

OHSSs may play a role in reducing job stress and burnout in five ways:

- **Monitoring** By carrying out stress audits, personal screenings, and psychosocial check-ups, information is gathered that can be used for developing preventive programmes to counteract burnout.

- **Co-ordination and integration** Efforts of experts from various fields (occupational medicine, safety engineering, human factors, occupational psychology) are coordinated. In addition, different levels (i.e. individual, interface, organisational) and approaches (i.e. clinical and organisational) are integrated into a comprehensive programme (Corey and Wolf, 1992).

- **Providing counselling services** Specialised counselling may be offered to employees with work-related mental problems such as burnout. Recently, Leakey *et al.* (1994) reported on the institutionalisation of a staff counselling service that was initiated by the North Derbyshire Health Authority in the United Kingdom as a part of a series of initiatives under the title 'Caring for the Carers'. Cooper and Sadri (1995) evaluated the impact of the implementation of an individual client-centred stress counselling service for employees of the UK Post Office. They observed a significant improvement in unauthorised absence events among those who visited the counselling service (27%), against only a small, non-significant improvement in the matched control group (5%). In addition, compared with the non-treated control group the authors found a significant decrease in anxiety and depression as well as an increase in self-esteem for those who participated in the counselling programme. Levels of job satisfaction and organisational commitment did not change, though.

- **Referral and expert consultation** Based on a sound assessment of the employee's work-related problems, the OHSS may consider referring the employee to local or community (mental) health services. Managers, supervisors, or employees might like to consult OHSS-experts on stress-related issues like, for example, how to deal with burned-out colleagues or subordinates.

- **Rehabilitation** OHSS-experts are important agents in rehabilitating employees who are in the process of recovering from burnout. They play an important mediating role between the workplace and the health-care professionals who are involved in treating burned-out employees.

Employee Assistance Programmes

Originally, Employee Assistance Programmes (EAPs) focused on the alcoholic employee. More recently, their scope has broadened and may now include job stress and burnout as follows from the American Employee Assistance Professionals Organisation's definition:

> An EAP is a worksite-based program to assist in the identification and resolution of productivity problems associated with employees impaired by personal concerns including, but not limited to: health, marital, family, financial, alcohol, drug, legal, emotional, stress, or other personal concerns which may adversely affect employee job performance.   (Lee and Gray, 1994, p. 216)

The ultimate concern of EAPs is with prevention, identification, and treating personal problems that adversely affect job performance. Hence, EAPs are more comprehensive than counselling services which primarily focus on treatment. To our knowledge, no EAPs exist that specifically focus on reducing burnout.

Since EAPs are meant to increase job performance they have been evaluated primarily in terms of cost-effectiveness. For instance, the US Department of Labor (1990) has calculated that for every $1 invested in an EAP, savings of $5 to as much as $16 are achieved. This makes EAPs quite attractive to employers. Yet, despite the dramatic expansion in the implementation of EAPs in the United States – about 12–15% of the American workforce is offered EAPs (Corey and Wolf, 1992) – they received little attention in Europe. Probably, EAPs do not fit into the European occupational welfare tradition because their focus is on adapting the employee instead of improving the working conditions or the work environment.

Outplacement

Organisations may offer employees outplacement services when it is most likely that successful rehabilitation can only be achieved in another job outside that organisation. Usually, outplacement is the outcome of a careful self-analysis and opportunity analysis that is carried out as part of a career development process (see above). Traditionally, outplacement is used for (top-)managers, but because of high financial risks (law suits) it has become increasingly attractive for professionals, such as physicians, nurses, and police officers.

6.5 BURNOUT WORKSHOPS

Some interventions that were discussed in the previous three sections are combined in burnout-workshops. Although such workshops are a booming business, their popularity stands in sharp contrast to the dearth of literature and research available on the subject. Burnout workshops are only occasionally described in greater detail and research on their effectiveness is even more scarce. In addition, most workshops lack a theoretical rationale. This is not so surprising because the field is dominated by practitioners whose interests are primarily commercial rather than scientific.

In Box 6.8 a detailed example is presented of a burnout workshop that has been conducted hundreds of times in the United States as well as in other countries with such different professional groups as day-care workers, doctors, dentists, teachers, university staff, nurses, correctional officers, telephone operators, and managers.

The workshop rests on two pillars: increasing the participants' awareness of their work-related problems and augmenting their coping resources by cognitive and behavioural skills training and by establishing support networks. More specifically, the workshop includes self-assessment, didactic stress-management, relaxation, cognitive and behavioural techniques, time-management, and peer support. Moreover, a more realistic image of the job is promoted and some organisational based interventions are discussed (e.g. how to improve job content, time scheduling, and communication). In other words, the burnout workshop combines many rather general strategies for one specific purpose: to prevent and combat burnout.

Effectiveness of burnout workshops

Do burnout workshops work? This question is rather difficult to answer because, as noted before, properly designed studies on the effectiveness of multifaceted burnout workshops are rare. Pines and Aronson (1983) evaluated a brief 1 day workshop for employees of two social services that broadly followed the steps outlined in Box 6.8. Although the authors claim that at the 1 week follow-up the level of burnout (i.e. exhaustion) had decreased slightly in the experimental group, this effect did not quite reach statistical significance. However, compared with the control group that did not participate in the workshop, satisfaction with colleagues and supervisors went up in the experimental group. The fact that participants are more satisfied than non-participants is in agreement with results from other evaluation studies on staff-support groups (cf. Brown, 1984; Larson, 1986). In addition, Pines and Aronson (1983) observed that awareness of the relationship between various work features and the experience of burnout was increased among the participants of the workshop. Unfortunately, due to the high attrition rate of participants in this study, the long term effects after 6 months could not be evaluated.

Box 6.8   Example of a burnout workshop

Step 1: Introducing oneself   Each participant introduces him- or herself by saying something personal about their jobs and the problems they were facing. Soon after the first participants have told their stories, it becomes clear that everybody in the group is struggling with similar problems. The awareness that problems are shared is important for counteracting both social isolation and feelings of false uniqueness.

Step 2: 'Breaking the ice'   Participants are asked in small groups to describe those qualities that friends and colleagues like and admire most about them. In this way, the participants are given the opportunity to boast in a socially accepted way. This has a two-sided effect: not only does it provide the group members with personal information of others, but it also makes participants feel more at ease.

Step 3: Self-assessment   The first part of the workshop is concluded with self-assessment consisting of a burnout questionnaire and a number of open questions about initial job expectations, work stressors, and ways of coping with it. Answering these questions further increases the participant's self-awareness.

Step 4: Awareness of the causes of burnout   Each participant is asked to describe to the other members of the small group what his or her goals and expectations were when (s)he entered the current job. Each group then chooses a goal or an expectation they all share to present to the rest of the larger group. The shared goals and expectations are written on the blackboard. The same procedure is repeated, but now for current burnout-causing stressors. The shared stressors are also written on the blackboard. And what is the result? The burnout-causing stressors can be stated, in almost every case, as frustrated goals and expectations! For instance, social workers find most stressful the bureaucratic hassles that interfere with adequately helping clients who are in need, and teachers find most stressful the discipline problems that make it difficult for them to educate and inspire students.

Step 5: Awareness of coping techniques   The realisation that current stressors are in fact frustrated goals and expectations is used as a lead-in to a discussion on coping. In a similar way as in the previous step, participants discuss in small groups how they cope with stressors at work. The most successful ways of coping are written on the blackboard. A plenary group discussion follows about the effectiveness of the various coping strategies.

Step 6: Coping skills training   The participants are taught various coping strategies such as relaxation exercises, time-management, or decompression. In addition, organisational coping strategies are discussed, such as making 'time outs' available, initiating discussion of disruptive emotional experiences, improving communication, rotating stressful jobs, or reducing bureaucratic interference.

Step 7: Awareness of social support The participants review their own lives and jobs to see which support functions are fulfilled for them and which are not. They present their conclusions to the other members of the small group. Next, the participants are asked to concentrate on those aspects that are not adequately fulfilled and to consider all the people they know who may play a role, and how they can ask them for help. The other members of the small group are asked to serve as challengers, forcing the person to examine excuses and make plans.

Step 8: Planning the future Using guided imagery participants are led to imagine, in as detailed a manner as possible, a typical day in their lives 5 years into the future. Next, based on their projections of the future, participants are asked to make a concrete plan that will it make more likely that the future unfolds in the way they envisioned it. By the end of the workshop the small group has developed into a closely knit social support network that might continue to function after the workshop is closed. The workshop ends with final feedback, leave-taking and a date for a follow-up meeting after a couple of months. At that meeting the participants' plans will be reviewed.

Source: Pines *et al.* (1981); Pines and Aronson (1983; 1988)

Schaufeli (1995) evaluated a somewhat similar 3 day workshop that was conducted for community nurses. The workshop included relaxation training, didactic stress management, RET, interpersonal skills training, and the enhancement of more realistic professional role expectations. At the 1 month follow-up the nurses' symptom levels (i.e. emotional exhaustion, psychological strain, and somatic complaints) decreased significantly. No significant changes were observed in levels of depersonalisation or personal accomplishment. Furthermore, it was observed that nurses who were more resistant to stress benefited more from the workshop than their colleagues who were less resistant. Unfortunately, this study did not include a control group.

Van Dierendonck *et al.* (in press) evaluated a 5 day burnout workshop for staff working in direct care with the mentally disabled, using a more rigorous design. The workshop was strongly cognitive-behaviourally oriented and included such aspects as cognitive restructuring, didactic stress management, and relaxation. In addition, a strong accent was put on career development. Like the workshop described in Box 6.8, initial goals and expectations were discussed and participants were encouraged to look for opportunities for personal growth in their current jobs. Furthermore, participants analysed their strengths and weaknesses and drew up a plan of action for the time ahead. This plan was directed either towards changing particular aspects of their current job, or looking for another job that would be more in line with their present goals and expectations. After 6 month and 1 year follow-up meetings were organised in order to evaluate the plan. The study included two non-treated control groups, one from the same organisation (the internal control group) and one from another similar organisation (the external control group). Results showed that emotional exhaustion dropped significantly for the experimental group compared with both control groups at each follow-up. Contrary to expectations, personal accomplishment had decreased in the experimental group at the 6 month follow-up compared with the external control group. However, this effect disappeared after another 6 months. Since no difference with the internal control group was observed in personal accomplishment organisation-specific effects (e.g. increased workload) could not be ruled out. No effects were observed for depersonalisation. Finally, registered absenteeism significantly decreased in the experimental group, whereas it increased in the internal control group.

Finally, Enzmann et al. (1992) evaluated a 3 day burnout workshop for human services professionals (mostly hospice staff) that was spread across 3 weekly intervals. The participants were encouraged to keep a stress diary, to use their newly developed cognitive and behavioural skills during the weekly intervals, and to report on their last week's experiences in the next session. The workshop included didactic stress management, relaxation training, coping skills training, interpersonal skills training to foster a professional attitude of detached concern, and sought to improve social support. At the 2 month follow-up levels of emotional exhaustion were significantly lower in the experimental group compared with the non-treated control group. No significant effects on the other two MBI burnout-dimensions were observed.

Taken together, it seems that multifaceted burnout workshops are effective in reducing levels of emotional exhaustion, even across relatively long periods of time (up to 1 year). Other burnout dimensions are usually not affected, though. These results are in agreement with previously discussed studies that investigated the burnout reducing effect of particular individual strategies (Corcoran and Bryce, 1983; West et al., 1984; Higgins, 1986; Freedy and Hobfoll, 1994). As a rule, these studies report positive effects on emotional exhaustion, but not on either of the other MBI burnout-dimensions. Most strategies that are evaluated in these studies are cognitive-behavioural in nature (e.g. relaxation, RET, SIT, time-management, didactic stress management, and cognitive restructuring). Accordingly, it is likely that this type of strategy is responsible for the reduction in exhaustion levels, whereas other strategies, such as enhancing social support, seem not to be effective. The fact that depersonalisation and reduced personal accomplishment do not change as a result of the workshop is not very surprising because most techniques employed in the workshops focus on reducing arousal and not on changing attitudes (depersonalisation) or on enhancing specific professional skills or resources (personal accomplishment).

6.6 SUMMARY

In this chapter we reviewed more than 30 different approaches to preventing or to combating burnout, and yet, there appears to be no general recipe. Thus, the reader is probably left with mixed feelings. On the one hand, there is some reason for pessimism. For instance, most interventions are rather general and are not specifically tailored to reduce burnout. Furthermore, there are only a few well-designed studies that document the effectiveness of interventions. Also, the focus of the interventions is rather biased towards the individual and organisational based interventions are relatively scarce. Viewed from another perspective, though, there is some reason for optimism. There is a vast and differentiated array of interventions available that may potentially be used to reduce burnout. These interventions differ in level (individual, interface, and organisational) and in purpose (ranging from identification, via prevention and treatment to rehabilitation). This broad variety is necessary because burnout is a complex phenomenon that may have many causes and that may develop along many different lines. Depending on the nature and the stage of the process and on the specific context in which it develops, different interventions may be used. So, at least there is some choice in measures to be taken (see Table 6.1). Moreover, a few strategies have proven to be effective in reducing burnout – another reason for some optimism. For instance, individual based cognitive and behavioural strategies, which are usually included in burnout workshops, appear to reduce burnout, most notably emotional exhaustion.

What can be said about the necessary ingredients of a burnout prevention programme? Three elements seem to be important for interventions at the individual as well as at the organisational level. First, levels of burnout have to be assessed so that the individual's or the organisation's awareness of the problem is increased. Second, measures should be taken to reduce negative arousal. At the individual level this can be achieved by using cognitive-behavioural techniques and at the organisational level by removing or reducing work stressors. Third, the person-job fit should be improved. At the individual level, skills must be enhanced so that employees are less vulnerable and more stress resilient (e.g. time-management) and more realistic expectations of the job must be promoted. At the organisational level, particular aspects of the job must be changed in order to better meet the needs of the employees (e.g. time scheduling).

CHAPTER SEVEN

Quo vadis?

Remaining issues

In recent decades numerous scholars and practitioners in many countries around the globe have contributed to our knowledge of burnout. Yet, some issues are still hotly debated or remain unresolved. In this concluding chapter we briefly discuss nine remaining issues and indicate how burnout researchers and practitioners may continue their path.

The popularity trap

Burnout has a strong public appeal and this is likely to remain the case in the years to come. It is still a pressing social problem that flourishes in today's modern society because of particular social, cultural, and economic developments such as the rapid growth of the service industry, increasing emotional workload, the individualisation of society, and the violation of the psychological contract between the employer and the employee. In many jobs and organisations, burnout has become a major expense in terms of absenteeism, reduced quality of service, and poor performance. Although hard scientific evidence for the organisational costs of burnout is rather scarce and equivocal (see Chapter 4), there can be few doubts about the significance of burnout for organisational life. Therefore, in the 1990s, burnout has appeared more and more on the agenda of many organisations, not just because of financial reasons, but also because of legal pressure. In many countries, particularly in Europe, legislation on occupational health, safety and well-being coerce employers to actively pursue the prevention of job stress and burnout.

At the same time as burnout receives attention in the workplace, it is treated as a hot item in the mass media. On the positive side, the great popularity of burnout fosters its recognition and creates a fertile soil for research and for institutionalising intervention programmes. On the negative side, however, there is the real danger of the 'popularity trap'. That is, the popularity of burnout stimulates the articulation of quick and simple solutions, dubious assessment methods, and inferior interventions by those who want to make fast money in the booming burnout business. Inevitably, the commercialisation of the popular burnout concept takes its toll – it remains associated with myths, fairy tales, and unverified 'facts'.

'Easy to see but hard to define'

Burnout is a powerful metaphor that is easily understood, applied, and recognised. Yet, it is difficult to define. In this respect burnout can be compared with pornography: nobody can define it but everybody recognises it instantly! Most coworkers, supervisors, managers, and executives have little difficulty in identifying 'burned-out' employees but at the same time they find it utterly difficult to come up with a concise definition.

As we have seen, it is not possible to define burnout in terms of its specific symptoms because of their sheer number – we identified over 130! Since an inductive laundry list approach is not effective in defining burnout, a deductive approach would be more appropriate. Yet, the most widely used instrument to measure burnout, the MBI, has been constructed inductively by analysing a set of items that were assumed to represent specific burnout symptoms (see below). Although the psychometric features of the MBI are satisfactory, a considerable overlap exists with depression, negative affectivity, and psychosomatic complaints. This lack of specificity of the MBI follows from the inductive approach that guided its construction. Using such an approach, any list of specific burnout symptoms will necessarily include symptoms that are also found in other syndromes because burnout – like other syndromes such as depression – cannot be appropriately defined solely at the level of specific symptoms. So, quite paradoxically, any burnout instrument that is based on specific symptoms tends to be rather unspecific because it will necessarily overlap with measures that tap related psychological conditions.

Alternatively, in our working definition, we characterised burnout in terms of general psychological constructs that cannot be observed directly (e.g. exhaustion, decreased motivation, reduced effectiveness, and the development of dysfunctional attitudes and behaviours). Our working definition deduces crucial common elements from many other definitions of burnout and thus acts as a common denominator. In addition, the definition includes a notion on the development of burnout: it is assumed that burnout results from a misfit between intentions and reality at the job, and that it is maintained through inadequate coping strategies. Unlike an inductive approach in defining burnout, a deductive approach remains necessarily broad and abstract. Hence, it must be concluded that, quite paradoxically, the specificity of burnout is best reflected by using an abstract definition instead of merely listing its symptoms.

The fact that it is so hard to define what burnout *really* is may be due to its various faces. It is most likely that the concept of burnout, as it is presently used, includes different subtypes. For instance, Gillespie (1981) distinguished between active and passive burnout. The former refers to assertive behaviour, whereas the latter is characterised by withdrawal and apathy. Moreover, burnout is usually associated with overload and energy depletion due to excessive job demands that is reinforced by high internal demands. For instance, 'early career burnout' is assumed to result from reality shock: a mismatch between everyday reality at the job and the newcomer's expectations (Cherniss, 1995). However, by contrast, burnout has also been described as 'worn out' (Cox *et al.*, 1993) or as 'tedium' (Pines *et al.*, 1981), which refers to underload and lack of challenge rather than to overload. Possibly, this type of burnout characterises those who are in more advanced stages of their careers. Based on our review of theoretical approaches, a third type of burnout may be distinguished that results from the blocking of work goals that are central to the person's professional identity. A prototypical example is the social worker who spends too much time filling out forms instead of actually helping people in need. Thus, it seems that different pathways lead to a similar negative mental state that is labelled 'burnout'. In order to understand the nature of burnout it is important to discriminate between such pathways.

Diagnostic or diabolic label?

Opinions on the use of burnout as an individual diagnostic label differ greatly. For some, it is a diabolic label because it may lead to the stigmatising of employees and increases the risk of individualisation and medicalisation. For them, using 'burnout' as a diagnostic label implies that it is a clinical syndrome instead of an occupational problem for which not the individual but the organisation is primarily responsible. Maslach and Leiter (1997, p. 34) strongly warn against this tendency that easily leads to blaming the victim:

> Burnout does not result from a genetic predisposition to grumpiness, a depressive personality, or general weakness. It is not caused by a failure of character or a lack of ambition. It is not a personality defect or a clinical syndrome. It is an occupational problem.

By contrast, we argue that burnout is both an occupational problem and a clinical syndrome. That is, burnout is an occupational problem that must be tackled by organisation-based preventive strategies as well as an individual problem that must be tackled by individual based treatment. The appropriateness of the perspective largely depends on the severity of burnout. When burnout symptoms are less severe it is most appropriate to consider it an occupational problem that can be solved by taking organisational measures. Paine (1982, p. 6) called this the Burnout Stress Syndrome:

> . . . the identifiable clusters of feelings and behaviours most commonly found in stressful or highly frustrating work environments.

By contrast, in cases of severe clinical forms of burnout it is more appropriate to focus on the person and on individual treatment. In terms of Paine (1982, p. 6) we are dealing with Burnout Mental Disability:

. . . the often serious clinically significant pattern of personal distress and diminished performance which is an end state of the burnout process.

He emphasises that the Burnout Stress Syndrome is not typically a mental disorder, but that it may lead to a possible disorder, the putative Burnout Mental Disability, which requires intensive psychological treatment.

Instead of Paine's Burnout Mental Disability, we prefer to use the term work-related neurasthenia because it refers to an existing diagnostic label that is based on the ICD-10 diagnostic guidelines. Of course, we realise that the use of a formal psychiatric label increases the risk of stigmatisation, individualisation, and medicalisation and could therefore be dismissed as a diabolic label. Yet, we believe that a formal recognition in terms of an official medical diagnosis is crucially important since it is the 'entry ticket' to our society's legal, health care, and social welfare systems. For workers to pursue their disability claims or to seek professional treatment, a sanctioned medical diagnosis is a necessary formal requirement.

'Burnout is what the MBI measures'

Burnout research is clearly dominated by the MBI: over 90% of the studies used this instrument. Since the MBI is *the* instrument to assess burnout, the concept is in fact narrowed to the three dimensions it includes – emotional exhaustion, depersonalisation, and reduced personal accomplishment. The advantage of this common standard is that research findings from different studies can be compared straightforwardly, as is shown in Chapters 3 and 4. However, the price to be paid for this uniformity is the rather narrow focus: burnout is what the MBI measures. This tautology is a serious problem since, as we noted above, the MBI has been developed inductively by factor-analysing a rather arbitrary set of items. What would have happened if other items had been included? Most likely, other dimensions would have appeared! In fact, various definitions and theoretical approaches suggest that the concept of burnout is much broader and more comprehensive than the MBI assumes. Because of the narrow scope of the MBI, a discrepancy exists between the richness and complexity of many conceptualisations of burnout, and the relatively simple nature of most empirical work. It remains a future challenge to develop an instrument to assess burnout that is based on a thorough theoretical analysis. Our working definition and our integrative model may serve as a starting point.

Is burnout a helper syndrome?

Burnout has first and foremost been studied in the human services. But is it really a helper syndrome, or may it also occur in other occupational fields? From the outset it has been claimed by practitioners and prominent scholars alike that burnout is not restricted to the human services. However, because of the almost universal acceptance of the MBI as the gold standard to measure burnout, the tendency to consider it as a specific helper syndrome has been affirmed – burnout is what the MBI measures *in the human services*. Attempts to use the MBI in non-human services settings have been criticised because the assumed equivalence with the original measure is questionable. Besides, it may be doubted on theoretical grounds whether burnout manifests itself in a similar fashion across all occupations. For instance, it is plausible that depersonalisation is a specific form of mental distancing that is typical for burnout in the human services, whereas in other occupations mental distancing may be expressed differently by, for example, being cynical about one's work. Based on this reasoning that asserts that the MBI-dimensions are specific expressions of a general burnout syndrome that also occurs outside the human services, the MBI-General Version was developed. With this instrument, burnout can be investigated in every occupational setting. On the one hand, this development should be welcomed because it enriches traditional job stress research by including in addition to an orthodox stress reaction (exhaustion), two other evaluative aspects that have not been extensively studied so far (cynicism and lack of professional efficacy). On the other hand, from a theoretical point of view, this development can be argued against. By generalising burnout beyond the human services it is deprived of its specific interpersonal nature.

The burnout epidemic: fact or fiction?

We know relatively little about the prevalence of burnout. Our present knowledge on levels of burnout across occupations is largely based on secondary analyses. The reason for this obvious omission is twofold. First, no large random, representative samples exist that allow a reliable estimate of the prevalence of burnout to be established. Secondly, and even more importantly, no valid cut-offs are available that discriminate burned-out from non-burned-out employees. In other words, what MBI-score indicates that an employee is burned-out? – provided that the MBI is a valid indicator. To our knowledge, except for the Dutch version of the MBI no clinically validated cut-offs are available. Based on these preliminary cut-offs, between 3% and 16% of Dutch workers suffer from clinical burnout syndrome depending on the respective MBI-dimension (Schaufeli and Van Dierendonck, 1995). These figures stand in contrast to popular opinion and to the claims of other researchers who conclude that 'Burnout seems epidemic, at least' (Golembiewski *et al.*, 1996, p. 163). We believe, however, that this conclusion is premature and that further research using representative samples and clinically valid cut-offs is needed.

Yet, a comparison of relative burnout levels across various occupations revealed interesting occupation-specific profiles that seem to be consistent across the two countries with the largest MBI data-bases: the United States and The Netherlands (see Chapter 3). For instance, in teaching, relatively high levels of emotional exhaustion are found; in the social services, levels of depersonalisation and reduced personal accomplishment are particularly high; and law enforcement (police and prisons) is characterised by low levels of emotional exhaustion and high levels of both other burnout dimensions.

Lots of data . . . but little knowledge

Relative to the huge amount of empirical studies (over 500!), our knowledge about causes and consequences of burnout as well as about the underlying psychological mechanism is still relatively poor. The reason is quite simple: most studies lack a firm theoretical framework and they are not adequately designed to deal with the issue of causality. Surprisingly few sound longitudinal studies have been conducted and, what is more, these studies do not unequivocally confirm the results from cross-sectional studies. Hence, as far as burnout is concerned, no definite conclusions about cause-effect relationships can be drawn.

This somewhat disappointing state of affairs is not only due to the poor quality and quantity of longitudinal research, but also to the high stability of burnout across time. The fact that burnout scores correlate highly across time may either mean that burnout is a trait-like mental condition or that the stressful job environment remains stable over time. In principle, longitudinal research in rapidly changing work environments (i.e. downsizing, introduction of new technologies) or among newcomers who enter their careers might solve this issue. However, there is also a more basic problem at stake here. It is most likely that burnout is a long-term stress reaction which means that, by definition, a considerable time interval exists between the exposure to causal agents and the emergence of symptoms. Moreover, the development of burnout is a dynamic process. That is, coping and adaptation play a key role. As a result of its chronic and dynamic nature it is virtually impossible to single out particular 'causes' that can be held responsible for the existence of what has become a generalised, chronic and habitual pattern of symptoms, labelled burnout. Viewed from this perspective it would be impossible to unravel the causes of burnout, even with sound longitudinal research. A possible way out could be to study a small sample very intensively over a relatively lengthy period of time. The disadvantage of this approach is that conclusions are difficult to generalise beyond the particular sample under investigation.

Although no causal inferences can be made from cross-sectional research it is not worthless, of course: it tells us something about the concomitants and correlates of burnout. High correlations with personality traits are noteworthy, especially neuroticism and negative affectivity, and with depression and self-reported psychosomatic complaints. Furthermore, burnout is related to quantitative workload instead of qualitative demands that result from problems in interacting with difficult recipients or dealing with their emotional problems. This is quite remarkable because traditional conceptualisations of burnout assume that emotionally charged interactions with recipients are its root cause. This assumption is not clearly confirmed by empirical research, though. Viewed from this perspective it is not surprising that recently the advocates of the influential emotional overload model changed their opinion and now consider emotional overload as only one of the possible causes of burnout that is embedded in the larger context of person-job mismatches (Maslach and Leiter, 1997).

Because most studies are exclusively based on self-reports there is a danger of falling into the so-called 'triviality trap'. This is illustrated by the fact that burnout tends to be much more strongly related to all kinds of self-reports than to more or less objective indicators of workload, performance, and absenteeism. Nevertheless, there is a great need to further explore relationships with such objective measures because of their crucial relevance to organisational life.

What makes burnout something special?

For at least two reasons, we believe, burnout is special. First, a typical pattern of emotional and cognitive processing is observed in which at least four psychological principles play a role that constitutes the uniqueness of the syndrome:

- **Meaning** Employees burn out only when their deeply rooted personal and professional motivations are challenged. Unmet expectations or frustrated intentions as such are not responsible for burnout. Only highly valued outcomes that are essential for one's professional identity may produce burnout because they are connected with existential expectations of significance, purpose and meaning.

- **Reciprocity** It is not the mere overload – working too hard, too long with too difficult recipients – that causes burnout but lack of reciprocity. That is, the imbalance between efforts and rewards. A strenuous job does not produce burnout as long as objective (e.g. pay, spare time) and subjective (e.g. gratitude, fulfilment) outcomes match the employee's efforts.

- **Unconscious processes** It seems that unconscious processes such as narcissistic strivings and the dynamics of conscious and unconscious psychic functions may contribute to burnout. This might explain why the process of burnout may remain unnoticed for a long time by the individual involved. After all, denial and suppression are instrumental for maintaining the individual's psychological integrity.

- **Personality** The development of burnout seems to be moderated by relatively stable individual characteristics that influence social information processing such as communal orientation, need for social comparison, career orientation, and 'feeling type'.

Second, burnout emerges in a social context that consists of three, mutually interacting levels:

- **Interpersonal** Burnout has to be understood first and foremost within the interpersonal context of the job that not only includes relationships with recipients but also with colleagues and supervisors. Social comparison and social induction processes, including emotional contagion, are likely to play a role in the development of burnout.

- **Organisational** Not only, as initially assumed, the interpersonal relationship with others at work, but also the relationship with the organisation is important for understanding burnout. The organisation imposes specific job demands (that may be too high), provides resources (that may be too few or not appropriate), and sets rules on how to express emotions (that may be too hard to comply with). On the subjective level the exchange relationship with the organisation is reflected in the employee's psychological contract. Accordingly, this contract is likely to play a crucial rule in the development of burnout.

■ **Societal** Burnout is not an exclusively subjective phenomenon, it has to be understood within a broader societal and cultural context. Developments on these levels, which are beyond the individual's control, influence the burnout process (e.g. the tendency for objectification; see Section 5.4). However, despite this, employees are not merely passive victims but are active agents. Individually as well as collectively they can, albeit within certain limits, shape their own conditions in such a way as to counteract burnout.

Unspecific interventions for a specific problem

'Nothing is as practical as a good theory' as Kurt Lewin (1945, p. 129) has stated. But unfortunately there is no 'good theory' of burnout, and there probably never will be. Nevertheless, a strong need remains to combat burnout. This dilemma is solved by applying the usual stress management strategies. Consequently, most burnout interventions lack specificity. That is, they focus on general job stressors and on general job-related tension rather than on specific causes and specific symptoms of burnout. This tendency is not as problematic as it may look at first glance, though. First, many presumed causes of burnout are quite general in nature, for example work overload and role problems. Second, emotional exhaustion in particular is related to other negative symptoms such as depressed mood and psychosomatic complaints. Nevertheless, more specific approaches are needed that address the distinctive features of burnout that were outlined above.

FINAL REMARK

The purpose of this book was to present a state-of-the-art overview of over two decades of scholarly and practical concern with the phenomenon of burnout. However, it was also written to act as an antidote against commercialised, ideological, and uncritical views on burnout. We attempted to demystify burnout by emphasising throughout the book the importance of evidence-based knowledge. In doing so we have probably raised more questions than we were able to answer. However, we strongly believe that, at the end of the day, a critical analysis of theory, research, assessment, and interventions is much more fruitful than an uncritical derivative or imitation of popular unverified views on burnout.

References

ABBOTT, A. (1990) Positivism and interpretation in sociology: Lessons for sociologists from the history of stress research. *Sociological Forum*, **5**, 435–458.

ABERNETHY, A.D. (1995) The development of an anger management training program for law enforcement personnel. In *Job Stress Interventions* ed. L.W. MURPHY, J.J. HURRELL, S.L. SAUTER and G.P. KEITA, pp. 21–30, Washington, DC: American Psychological Association.

ACHTERHUIS, H. (1979) *De markt van welzijn en geluk* [The market of wellbeing and happiness]. Baarn: Ambo.

ACKERLEY, G.D., BURNELL, J., HOLDER, D.C. and KURDEK, L.A. (1988) Burnout among licensed psychologists. *Professional Psychology: Research and Practice*, **19**, 624–631.

AMERICAN PSYCHIATRIC ASSOCIATION (1994) *Diagnostic and Statistical Manual of Mental Disorders*, 4th edn., Washington, DC: American Psychiatric Association.

APPELS, A. and SCHOUTEN, E. (1991) Burnout as a risk factor for coronary heart disease. *Behavioral Medicine*, **17**, 53–59.

ASHFORTH, B.E. and HUMPHRY, R.H. (1993) Emotional labour in service roles: The influence of identity. *Academy of Management Review*, **18**, 88–115.

BAKKER, A.B., KILLMER, C., SIEGRIST, J. and SCHAUFELI, W.B. (1998a) *Effort-reward imbalance and burnout among nurses* (Unpublished manuscript, Psychological Department, Utrecht University).

BAKKER, A.B., LE BLANC, P.M. and SCHAUFELI, W.B. (1998b) *Burnout contagion: Are ICU-nurses at risk for infection with the burnout virus?* (Unpublished manuscript, Psychological Department, Utrecht University).

BAKKER, A.B., SCHAUFELI, W.B., SIXMA, H., BOSVELD, W. and VAN DIERENDONCK, D. (1997) *Harassment by patients, perceptions of inequity, and burnout: A five-year longitudinal study among general practitioners* (Unpublished manuscript, Psychological Department, Utrecht University).

BANDURA, A. (1997) *Self-efficacy: the exercise of control.* New York: Freeman.

BARLEY, S.R. and KNIGHT, D.B. (1992) Toward a cultural theory of stress complaints. *Research in Organizational Behavior*, **14**, 1–48.

BATMAN, D.C. (1994) Development of a corporate wellness programme Nestlé UK Ltd. In *Creating Healthy Work Organisations* ed. C.L. COOPER and S. WILLIAMS, pp. 25–48, Chichester: John Wiley.

BEEMSTERBOER, J. and BAUM, B.H. (1984) 'Burnout': Definitions and health care management. *Social Work in Health Care*, **10**, 97–109.

BEER, J. and BEER, J. (1992) Burnout and stress, depression and self-esteem of teachers. *Psychological Reports*, **71**, 1331–1336.

BERG, A., WELANDER-HANSSON, U. and HALLBERG, I.R. (1994) Nurses' creativity, tedium and burnout during 1 year of clinical supervision and implementation of individually planned nursing care: Comparisons between a ward for severely demented patients and a similar control ward. *Journal of Advanced Nursing*, **20**, 742–749.

BHAGAT, R.S., ALLIE, S.M. and FORD, D.L. (1995) Coping with stressful life events: An empirical analysis. In *Occupational stress: A handbook. Series in health psychology and behavioral medicine* ed. R. CRANDALL and P.L. PERREWE, pp. 93–112, Philadelphia, PA: Taylor & Francis.

BIBEAU, G., DUSSAULT, G., LAROUCHE, L.M. *et al.* (1989) *Certain aspects culturels, diagnostiques et juridiques de burnout* [Some cultural, diagnostic and juridical aspects of burnout]. Montréal: Confédération des Syndicats Nationaux.

BIRCH, N.E., MARCHANT, M.P. and SMITH, N.M. (1986) Perceived role conflict, role ambiguity, and reference librarian burnout in public libraries. *Library & Information Science Research*, **8**, 53–65.

BLOSTEIN, S., ELDRIDGE, W., KILTY, K. and RICHARDSON, V. (1985) A multidimensional analysis of the concept of burnout. *Employee Assistance Quarterly*, **1**, 55–66.

BRADLEY, H.B. (1969) Community-based treatment for young adult offenders. *Crime and Delinquency*, **15**, 359–370.

BREAY, M. (1913) The overstrain of nurses. *The Canadian Nurse*, **9**, 153–157.

BRILL, P.L. (1984) The need for an operational definition of burnout. *Family and Community Health*, **6** (4), 12–24.

BROGMUS, G.E. (1996) The rise and fall? of mental stress claims in the USA. *Work & Stress*, **10**, 24–35.

BROOKINGS, J.B., BOLTON, B., BROWN, C.E. and McEVOY, A. (1985) Self-reported job burnout among female human service professionals. *Journal of Occupational Behaviour*, **6**, 143–150.

BROWN, L. (1984) Mutual help staff-groups to manage work stress. *Social Work with Groups*, **7**, 55–66.

BURISCH, M. (1989) *Das Burnout-Syndrom. Theorie der inneren Erschöpfung* [The Burnout Syndrome: A Theory of Inner Exhaustion]. Berlin: Springer.

BURISCH, M. (1993) In search of theory: Some ruminations on the nature and etiology of burnout. In *Professional Burnout: Recent Developments in Theory and Research* ed. W.B. SCHAUFELI, C. MASLACH and T. MAREK, pp. 75–93, Washington, DC: Taylor & Francis.

BURKE, J.M. (1985) The relationship between type A behavior, role stress, job enrichment and burnout among college counselors. *Dissertation Abstracts International*, **46**, 3588B.

BURKE, R.J. (1994) Stressful events, work-family conflict, coping, psychological burnout, and well-being among police officers. *Psychological Reports*, **75**, 787–800.

BURKE, R.J. and GREENGLASS, E.R. (1988) Career orientations and psychological burnout in teachers. *Psychological Reports*, **63**, 107–116.

BURKE, R.J. and GREENGLASS, E.R. (1989a) Correlates of psychological burnout phases among teachers. *Journal of Health and Human Resources Administration*, **12**, 46–62.

BURKE, R.J. and GREENGLASS, E.R. (1989b) Psychological burnout among men and women in teaching: An examination of the Cherniss model. *Human Relations*, **42**, 261–273.

BURKE, R.J. and GREENGLASS, E.R. (1991) A longitudinal study of progressive phases of psychological burnout. *Journal of Health and Human Resources Administration*, **13**, 390–408.

BURKE, R.J. and GREENGLASS, E.R. (1994) Towards an understanding of work satisfactions and emotional well-being of school-based educators. *Stress Medicine*, **10**, 177–184.

BURKE, R.J. and GREENGLASS, E.R. (1995) A longitudinal examination of the Cherniss model of psychological burnout. *Social Science and Medicine*, **40**, 1357–1363.

BURKE, R.J. and RICHARDSEN, A.M. (1993) Psychological burnout in organizations. In *Handbook of Organizational Consultation* ed. R.T. GOLOMBIEWSKI, pp. 263–298. New York: Marcel Dekker.

BURKE, R.J., SHEARER, J. and DESZCA, G. (1984) Burnout among men and women in police work: An examination of the Cherniss model. *Journal of Health and Human Resources Administration*, **7**, 162–188.

BÜSSING, A. and PERRAR, K.M. (1991) Burnout und Streß. Untersuchung zur Validität von Burnout und Streß in der Krankenpflege in Abhängigkeit von Geschlecht und beruflicher Position. [Burnout and stress. A study on the validity of burnout and stress in nursing as related to gender and occupational position]. In *Arbeitsbedingungen im Krankenhaus und Heim* [Working conditions in hospitals and hospices] ed. K. LANDAU, pp. 42–87, Stuttgart: Gentner.

BUUNK, B.P. and SCHAUFELI, W.B. (1993) Burnout: A perspective from social comparison theory. In *Professional Burnout: Recent Developments in Theory and Research* ed. W.B. SCHAUFELI, C. MASLACH and T. MAREK, pp. 53–69, Washington, DC: Taylor & Francis.

BUUNK, B.P. and SCHAUFELI, W.B. (1998) Reciprocity in interpersonal relationships: An evolutionary perspective on its importance for health and well-being. In *The European Review of Social Psychology* (Vol. 10) ed. W. STROEBE and M. HEWSTONE, Chichester: Wiley.

BUUNK, B.P., SCHAUFELI, W.B. and YBEMA, J.F. (1994) Burnout, uncertainty, and the desire for social comparison among nurses. *Journal of Applied Social Psychology*, **24**, 1701–1718.

BYRNE, B.M. (1994) Burnout: Testing for the validity, replication, and invariance of causal structure across elementary, intermediate, and secondary teachers. *American Educational Research Journal*, **31**, 645–673.

CALIFORNIA WORKERS' COMPENSATION INSTITUTE (1990) *CWCI Bulletin*. San Fransisco, CA: California Workers' Compensation Institute.

CAPEL, S.A. (1991) A longitudinal study of burnout in teachers. *British Journal of Educational Psychology*, **61**, 36–45.

CASH, D. (1988) A study of the relationship of demographics, personality, and role stress to burnout in intensive care unit nurses. *Dissertation Abstracts International*, **49**, 2585A.

CHERNISS, C. (1980a) *Professional Burnout in the Human Service Organizations*. New York: Praeger.

CHERNISS, C. (1980b) *Staff Burnout. Job Stress in the Human Services*. Beverly Hills, CA: Sage.

CHERNISS, C. (1989) Career stability in public service professionals: A longitudinal investigation based on biographical interviews. *American Journal of Community Psychology*, **17**, 399–422.

CHERNISS, C. (1990) Natural recovery from burnout: Results from a 10-year follow-up study. *Journal of Health and Human Resources Administration*, **13**, 132–154.

CHERNISS, C. (1992) Long-term consequences of burnout: An exploratory study. *Journal of Organizational Behavior*, **13** (1), 1–11.

CHERNISS, C. (1993) Role of professional self-efficacy in the etiology and amelioration of burnout. In *Professional Burnout: Recent Developments in Theory and Research* ed. W.B. SCHAUFELI, C. MASLACH and T. MAREK, pp. 135–149, Washington, DC: Taylor & Francis.

CHERNISS, C. (1995) *Beyond burnout: Helping teachers, nurses, therapists and lawyers recover from stress and disillusionment*. New York: Routledge.

CHERNISS, C. and DANZIG, S.A. (1986) Preventing and managing job related stress. In *Professionals in Distress: Issues, Syndromes and Solutions in Psychology* ed. R.R. KILBURG, P.E. NATHAN and R.W. THORESON, pp. 237–255, Washington, DC: American Psychological Association.

CHERRY, N. (1978) Stress, anxiety and work: A longitudinal study. *Journal of Occupational Psychology*, **51**, 257–259.

CLARK, L.A., WATSON, D. and MINEKA, S. (1994) Temperament, personality, and the mood and anxiety disorders. *Journal of Abnormal Psychology*, **103**, 103–116.

CLOUSE, R.W. (1982) *The Burnout Assessment Inventory*. Nashville, TN: Matrix Systems.

COHEN, J. (1977) *Statistical Power Analysis for the Behavioral Sciences* (Rev. edn). New York: Academic Press.

COHEN, M.E. and WHITE, P.D. (1951) Life situations, emotions and neurocirculatory asthenia. *Psychosomatic Medicine*, **13**, 335–355.

COLEGROVE, S.C.B. (1983) Personality and demographic characteristics as predictors of burnout in female police officers. *Dissertation Abstracts International*, **44**, 1232B.

CONLEY, J.J. (1984) The hierarchy of consistency: A review and model of longitudinal findings on adult individual differences in intelligence, personality and self-opinion. *Personality and Individual Differences*, **5**, 11–25.

CONNER, V.L. (1982) A comparative study of nursing burnout and somatic complaints in three occupational settings. *Dissertation Abstracts International*, **43** (6), 1887A.

COOLEY, E. and YOVANOFF, P. (1996) Supporting professionals-at-risk: Evaluating interventions to reduce burnout and improve retention of special educators. *Exceptional Children*, **62**, 336–355.

COOPER, C.L., LIUKKONEN, P. and CARTWRIGHT, S. (1996) *Stress prevention in the workplace: Assessing the costs and benefits to organisations*. Dublin: European Foundation for the Improvement of Living and Working Conditions, Loughlinstown House.

COOPER, C.L. and SADRI, G. (1995) The impact of stress counseling at work. In *Occupational Stress: A Handbook* ed. R. CRANDALL and P. PERREWÉ, pp. 271–282, London: Taylor & Francis.

COOPER, C.L., SLOAN, S.J. and WILLIAMS, S. (1988) *Occupational Stress Indicator Management Guide*. Windsor: NFER-Nelson.

CORCORAN, K.J. (1986) The association of burnout and social work practitioners' impressions of their clients: Empirical evidence. *Journal of Social Service Research*, **10**, 57–66.

CORCORAN, K.J. and BRYCE, A.K. (1983) Intervention in the experience of burnout: Effects of skill development. *Journal of Social Service Research*, **7**, 71–79.

CORDES, C.L. and DOUGHERTY, T.W. (1993) A review and an integration of research on job burnout. *Academy of Management Review*, **18**, 621–656.

CORDES, C.L., DOUGHERTY, T.W. and BLUM, M. (1997) Patterns of burnout among managers and professionals: a comparison of models. *Journal of Organizational Behavior*, **18**, 685–701.

COREY, D.M. and WOLF, G.D. (1992) An integrated approach to reducing stress injuries. In *Stress and Well-Being at Work: Assessments and Interventions for Occupational Health* ed. J.C. QUICK, R.L. MURPHY and J.J. HURRELL, pp. 64–78, Washington, DC: American Psychological Association.

CORRIGAN, P.W., HOLMES, E.P. and LUCHINS, D. (1995) Burnout and collegial support in state psychiatric hospital staff. *Journal of Clinical Psychology*, **51**, 703–710.

CORRIGAN, P.W., HOLMES, E.P., LUCHINS, D. *et al.* (1994) Staff burnout in a psychiatric hospital: A cross-lagged panel design. *Journal of Organizational Behavior*, **15**, 65–74.

COX, T. (1978) *Stress*. London: Macmillan.

COX, T., KUK, G. and LEITER, M.P. (1993) Burnout, health, work stress, and organizational healthiness. In *Professional Burnout: Recent Developments in Theory and Research* ed. W.B. SCHAUFELI, C. MASLACH and T. MAREK, pp. 177–193, Washington, DC: Taylor & Francis.

CUNNIGHAM, W.G. (1983) Teacher burnout – Solutions for the 1980's: A review of the literature. *The Urban Review*, **15**, 37–51.

DAMES, K.A. (1983) Relationship of burnout to personality and demographic traits in nurses. *Dissertation Abstracts International*, **44**, 1588B.

DE FRANK, R.S. and COOPER, C.L. (1987) Worksite stress management interventions: Their effectiveness and conceptualisation. *Journal of Managerial Psychology*, **2**, 4–10.

DE GIER, E. (1995) Occupational welfare in the European Community: Past, present and future. In *Job Stress Interventions* ed. L.W. MURPHY, J.J. HURRELL, S.L. SAUTER and G.P. KEITA, pp. 405–416, Washington, DC: American Psychological Association.

DEARY, I.J., BLENKIN, H., AGIUS, R.M. *et al.* (1996) Models of job-related stress and personal achievement among consultant doctors. *British Journal of Psychology*, **87**, 3–29.

DEKKER, S.W. and SCHAUFELI, W.B. (1995) The effects of job insecurity on psychological health and withdrawal: A longitudinal study. *Australian Psychologist*, **30**, 57–63.

DIGNAM, J.T. (1986) Social support, job stress, burnout and health among correctional officers: A longitudinal analysis. *Dissertation Abstracts International*, **47**, 4646B.

DIGNAM, J.T., BARRERA, M. and WEST, S.G. (1986) Occupational stress, social support, and burnout among correctional officers. *American Journal of Community Psychology*, **14**, 177–193.

DIGNAM, J.T. and WEST, S.G. (1988) Social support in the workplace: Tests of six theoretical models. *American Journal of Community Psychology*, **16**, 701–724.

DION, G. and TESSIER, R. (1994) Validation de la traduction de l'Inventaire d'épuisement professionnel de Maslach et Jackson. [Validation of the translation of the burnout inventory of Maslach and Jackson] *Canadian Journal of Behavioural Science*, **26** (2), 210–227.

DYMENT, W.E. (1989) Burnout among missionaries: An empirical inquiry into the role of unrealistic expectations, job role ambiguity and job role conflict. *Dissertation Abstracts International*, **50** (8), 3690B.

DYNERMAN, S.B. and HAYES, L.O. (1991) *The Best Jobs in America for Parents who want Careers and Time for Children too.* New York: Rawson.

EDELWICH, J. and BRODSKY, A. (1980) *Burn-Out: Stages of Disillusionment in the Helping Professions.* New York: Human Sciences Press.

EINSIEDEL, A. and TULLY, H. (1982) Methodological considerations in studying the burnout phenomenon. In *The Burnout Syndrome* ed. J.W. JONES, pp. 89–106, Park Ridge, IL: London House.

ELKIN, A.J. and ROSCH, P.J. (1990) Promoting mental health at the workplace: The prevention side of stress management. *Occupational Medicine: State of the Art Review*, **5**, 739–754.

ELLIS, A. (1962) *Reason and Emotion in Psychotherapy.* New York: Lyle Stuart.

EMENER, W.G., LUCK, R.S. and GOHS, F.X. (1982) A theoretical investigation of the construct burnout. *Journal of Rehabilitation Administration,* **6**, 188–196.

ENZMANN, D. (1996) *Gestreßt, erschöpft oder ausgebrannt? Einflüsse von Arbeitssituation, Empathie und Coping auf den Burnoutprozeß* [Stressed, exhausted, or burned out? Effects of working conditions, empathy, and coping on the development of burnout]. Munich: Profil.

ENZMANN, D., BERIEF, P., ENGELKAMP, C. *et al.* (1992) *Burnout & Burnoutbewältigung. Entwicklung und Evaluation eines Burnoutworkshops* [Burnout and coping with burnout. Development and evaluation of a burnout workshop] (Projektbericht des Studienprojekts). Berlin: Technische Universität Berlin, Institut für Psychologie.

ENZMANN, D. and KLEIBER, D. (1989) *Helfer-Leiden: Streß und Burnout in psychosozialen Berufen* [Helper-ordeals: Stress and burnout in the human services]. Heidelberg: Asanger.

ENZMANN, D., SCHAUFELI, W.B. and GIRAULT, N. (1995) The validity of the Maslach Burnout Inventory in three national samples. In *Health Workers and AIDS. Research, Intervention and Current Issues in Burnout and Response* ed. L. BENNETT, D. MILLER and M.W. ROSS, pp. 131–150, Chur (Switzerland): Harwood.

ENZMANN, D., SCHAUFELI, W.B., JANSSEN, P. and ROZEMAN, A. (1998) Dimensionality and validity of the Burnout Measure. *Journal of Occupational and Organizational Psychology.*

ETZION, D. (1987) *Burnout: The hidden agenda of human distress* (IIBR Series in Organizational Behavior and Human Resources, Working paper No. 930/87). Tel Aviv, Israel: The Israel Institute of Business Research, Faculty of Management, Tel Aviv University.

FARBER, B.A. (1983) Introduction: A critical perspective on burnout. In *Stress and Burnout in the Human Service Professions* ed. B.A. FARBER, pp. 1–20, New York: Pergamon.

FARBER, B.A. (1984). Stress and burnout in suburban teachers. *Journal of Educational Research,* **77**, 325–331.

FIELDING, J.E. (1989) Worksite stress management: National survey results. *Journal of Occupational Medicine,* **31**, 990–995.

FIGLEY, C.R. (ed.) (1995) *Compassion Fatigue: Coping with Secondary Traumatic Stress Disorder in Those who Treat the Traumatised.* New York: Brunner/Mazel.

FIMIAN, M.J. (1984) The development of an instrument to measure occupational stress in teachers: The Teacher Stress Inventory. *Journal of Occupational Psychology,* **57**, 277–293.

FINN, S.E. (1986) Stability of personality ratings over 30 years. Evidence of an age/cohort interaction. *Journal of Personality and Social Psychology,* **50**, 813–818.

FIRTH, H. and BRITTON, P.G. (1989) 'Burnout', absence and turnover amongst British nursing staff. *Journal of Occupational Psychology,* **62**, 55–59.

FISCHER, H.J. (1983) A psychoanalytic view of burnout. In *Stress and Burnout in the Human Service Professions* ed. B.A. FARBER, pp. 40–45, New York: Pergamon.

FONTANA, D. (1989) *Managing Stress.* London: Routledge.

FORD, D.L., MURPHY, C.J. and EDWARDS, K.L. (1983) Exploratory development and validation of a perceptual job burnout inventory: Comparison of corporate sector and human services professionals. *Psychological Reports,* **52**, 995–1006.

FREEDY, J.R. and HOBFOLL, S.E. (1994) Stress inoculation for reduction of burnout: A conservation of resources approach. *Anxiety, Stress and Coping,* **6**, 311–325.

FREUDENBERGER, H.J. (1974) Staff burn-out. *Journal of Social Issues,* **30**, 159–165.

FREUDENBERGER, H.J. (1983) Burnout: Contemporary issues, trends and concerns. In *Stress and Burnout in the Human Service Professions,* ed. B.A. FARBER, pp. 23–28, New York: Pergamon.

FREUDENBERGER, H.J., with RICHELSON, G. (1990) *Burn-Out: How to Beat the High Cost of Success.* New York: Bantam Books.

FRIESEN, D. and SARROS, J.C. (1989) Sources of burnout among educators. *Journal of Organizational Behavior,* **10**, 179–188.

GARDEN, A.M. (1985) The effect of Jungian type on burnout. *Journal of Psychological Type,* **10**, 3–10.

GARDEN, A.M. (1987) Depersonalization: A valid dimension of burnout? *Human Relations,* **40**, 545–560.

GARDEN, A.M. (1988) Jungian type, occupation and burnout: An elaboration of an earlier study. *Research in Psychological Type,* **14**, 2–14.

GARDEN, A.M. (1989) Burnout: The effect of psychological type on research findings. *Journal of Occupational Psychology,* **62**, 223–234.

GARDEN, A.M. (1991) The purpose of burnout: A Jungian interpretation. *Journal of Social Behavior and Personality,* **6**, 73–93.

GIBSON, F., McGRATH, A. and REID, N. (1989) Occupational stress in social work. *British Journal of Social Work,* **19** (1), 1–18.

GILLESPIE, D.F. (1981) Correlates for active and passive types of burnout. *Journal of Social Service Research,* **4** (2), 1–16.

GILLESPIE, D.F. (1983) *Understanding and combatting burnout* (Public Administration Series: Bibliography P-1235). Monticello, IL: Vance Bibliographies.

GIL-MONTE, P.R., PEIRÓ, J.M. and VALCÁRCEL, P. (1993) *Work overload, social support and self-confidence as burnout antecedents in nursing professionals* (Paper presented at the 3rd ENOP Workshop on Personnel Psychology in Health Care Organizations, September 23–25, 1993, Cracow (Poland)).

GLASS, D.C. and McKNIGHT, J.D. (1996) Perceived control, depressive symptomatology, and professional burnout: A review of the evidence. *Psychology and Health,* **11**, 23–48.

GLASS, D.C., McKNIGHT, J.D. and VALDIMARSDOTTIR, H. (1993) Depression, burnout and perceptions of control in hospital nurses. *Journal of Consulting and Clinical Psychology,* **61**, 147–155.

GLICKAUF-HUGHES, C. and MEHLMAN, E. (1995) Narcissistic issues in therapists: Diagnostic and treatment considerations. *Psychotherapy,* **32**, 213–221.

GOLD, Y., BACHELOR, P.A. and MICHAEL, W.B. (1989) The dimensionality of a modified form of the Maslach Burnout Inventory for university students in a teacher-training program. *Educational and Psychological Measurement,* **49**, 549–561.

GOLD, Y. and ROTH, R.A. (1993) *Teachers Managing Stress and Preventing Burnout.* London: The Falmer Press.

GOLEMBIEWSKI, G.T., BOUDREAU, R.A., MUNZENRIDER, R.F. and LUO, H. (1996) *Global Burnout: A Worldwide Pandemic explored by the Phase Model.* Greenwich, CT: JAI Press.

GOLEMBIEWSKI, R.T. and BOSS, R.W. (1991) Shelving levels of burnout for individuals in organizations: A note on the stability of phases. *Journal of Health and Human Resources Administration,* **13**, 409–420.

GOLEMBIEWSKI, R.T. and MUNZENRIDER, R.F. (1988) *Phases of Burnout: Developments in Concepts and Applications.* New York: Praeger.

GOLEMBIEWSKI, R.T., MUNZENRIDER, R.F., SCHERB, K. and BILLINGSLEY, W. (1992) Burnout and psychiatric 'cases': Early evidences of an association. *Anxiety, Stress and Coping,* **5**, 69–78.

GOLEMBIEWSKI, R.T., MUNZENRIDER, R.F. and STEVENSON, J.G. (1986) *Stress in Organizations: Toward a Phase Model of Burnout.* New York: Praeger.

GOLEMBIEWSKI, R.T. and ROUNTREE, B.H. (1991) Releasing human potential for collaboration: A social intervention targeting supervisory relationships and stress. *Public Administration Quarterly*, **15**, 32–35.

GOTTMAN, J.M. (ed.) (1995) *The Analysis of Change*. Mahwah, NJ: Lawrence Erlbaum.

GOW, K.M. (1982) *How Nurses' Emotions Affect Patient Care*. New York: Springer.

GRAHAM, S.D.J. (1993) Staff burnout and job-induced tension: The buffering effects of social support and locus of control. *Masters Abstracts International*, **33**, 371.

GRAY-TOFT, P. and ANDERSON, J.G. (1981) Stress among hospital nursing staff: Its causes and effects. *Social Science & Medicine*, **15**, 639–647.

GREENE, G.A. (1961) *A Burnt-out Case*. New York: Viking Press.

GROENESTIJN, E., BUUNK, A.P. and SCHAUFELI, W.B. (1992) Het besmettingsgevaar bij burnout: De rol van sociale vergelijkingsprocessen [The danger of burnout contagion: The role of social comparison processes]. In *Sociale Psychologie en Beïnvloeding van Intermenselijke en Gezondheidsproblemen* [Social psychology and influencing interpersonal and health problems] ed. R.W. MEERTENS, B.P. BUUNK, P.A.M. VAN LANGE and B. VERPLANKEN, pp. 88–103, The Hague: VUGA.

GUSY, B. (1995) *Stressoren in der Arbeit, Soziale Unterstützung und Burnout. Eine Kausalanalyse* [Job stress, social support, and burnout. A causal analysis] (Prävention und psychosoziale Gesundheitsforschung, Bd. 1). Munich: Profil.

GUY, J.D., POELSTRA, P.L. and STARK, M.J. (1989) Personal distress and therapeutic effectiveness: A national survey of psychologists practising psychotherapy. *Professional Psychology: Research and Practice*, **20**, 48–50.

HALLSTEN, L. (1993) Burning out: A framework. In *Professional Burnout: Recent Developments in Theory and Research* ed. W.B. SCHAUFELI, C. MASLACH and T. MAREK, pp. 95–113, Washington, DC: Taylor & Francis.

HAMANN, D.L., DAUGHERTY, E. and MILLS, C.R. (1987) An investigation of burnout assessment and potential job related variables among public school music educators. *Psychology of Music*, **15**, 128–140.

HAMMER, J.S., JONES, J.W., LYONS, J.S., SIXMITH, D. and AFFICIANDO, E. (1985) Measurement of occupational stress in hospital settings: Two validity studies of a measure of self-reported stress in medical emergency rooms. *General Hospital Psychiatry*, **7**, 156–162.

HANDY, J.A. (1988) Theoretical and methodological problems within occupational stress and burnout research. *Human Relations*, **41**, 351–369.

HANDY, J.A. (1991) Stress and contradiction in psychiatric nursing. *Human Relations*, **44**, 39–53.

HARATANI, T. (1997) Karoshi: Death from overwork. In *Encyclopaedia of Occupational Health and Safety* ed. INTERNATIONAL LABOUR ORGANISATION, pp. 518–519, Geneva: ILO.

HARE, J.M., PRATT, C.C. and ANDREWS, D. (1988) Predictors of burnout in professional and paraprofessional nurses working in hospitals and nursing homes. *International Journal of Nursing Studies*, **25**, 105–115.

HARRISON, W.D. (1983) A social competence model of burnout. In *Stress and Burnout in the Human Service Professions* ed. B.A. FARBER, pp. 29–39, New York: Pergamon.

HATFIELD, E., CACIOPPO, J.T. and RAPSON, R.L. (1994) *Emotional Contagion*. New York: Cambridge University Press.

HEDGES, L.V. and OLKIN, I. (1985) *Statistical Methods for Meta-Analysis*. Orlando, FL: Academic Press.

HEIFETZ, L.J. and BERSANI, H.A. (1983) Disrupting the pursuit of personal growth: Toward a unified theory of burnout in the human services. In *Stress and Burnout in the Human Service Professions* ed. B.A. FARBER, pp. 46–62, New York: Pergamon.

HENDRIX, W.H., STEEL, R.P., LEAP, T.L. and SUMMERS, T.P. (1991) Development of a stress-related health promotion model: Antecedents and organizational effectiveness outcomes. *Journal of Social Behavior and Personality*, **6**, 141–162.

HIGGINS, N.C. (1986) Occupational stress and working women: The effectiveness of two stress reduction programs. *Journal of Vocational Behavior*, **29**, 66–78.

HILLS, H. and NORVELL, N. (1991) An examination of hardiness and neuroticism as potential moderators of stress outcomes. *Behavioral Medicine*, **17**, 31–38.

HIMLE, D.P. and JAYARATNE, S.D. (1990) Burnout and job satisfaction: Their relationship to perceived competence and work stress among undergraduate and graduate social workers. *Journal of Sociology and Social Welfare*, **17** (4), 93–108.

HIMLE, D.P., JAYARATNE, S.D. and THYNESS, P.A. (1991) Buffering effects of four social support types on burnout among social workers. *Social Work Research and Abstracts*, **27**, 22–27.

HINGLEY, P. and COOPER, C.L. (1986) *Stress and the Nurse Manager.* London: John Wiley.

HOBFOLL, S.E. and FREEDY, J. (1993) Conservation of resources: A general stress theory applied to burnout. In *Professional Burnout: Recent Developments in Theory and Research* ed. W.B. SCHAUFELI, C. MASLACH and T. MAREK, pp. 115–129, Washington, DC: Taylor & Francis.

HOBFOLL, S.E. and SHIROM, A. (1993) Stress and burnout in the workplace: Conservation of resources. In *Handbook of Organizational Behavior* ed. R.T. GOLEMBIEWSKI, pp. 41–60, New York: M. Dekker.

HOCHSCHILD, A.R. (1983) *The Managed Heart. Commercialization of Human Feeling.* Berkeley, CA: University of California Press.

HOOGDUIN, C.A.L., SCHAAP, C.P.D.R., KLADLER, A.J. and HOOGDUIN, W.A. (eds) (1996a) *Behandelingsstrategieën bij burnout* [Burnout treatment programmes]. Houten: Bohn Stafleu van Loghum.

HOOGDUIN, W.A., VOSSEN, C. and HOOGDUIN, C.A.L. (1996b) Reïntegratie: Een geïntegreerde aanpak [Rehabilitation: An integrative approach]. In *Behandelingsstrategieën bij burnout* [Burnout treatment programmes] ed. C.A.L. HOOGDUIN, C.P.D.R. SCHAAP, A.J. KLADLER and W.A. HOOGDUIN, pp. 69–77, Houten: Bohn Stafleu van Loghum.

HOUTMAN, I.L.D. and KOMPIER, M.A.J. (1995) Risk factors and occupational risk groups for work stress in the Netherlands. In *Organizational Risk Factors for Job Stress* ed. S.L. SAUTER and L.R. MURPHY, pp. 209–226, Washington, DC: American Psychological Association.

HUBERTY, T.J. and HUEBNER, E.S. (1988) A national survey of burnout among school psychologists. *Psychology in the Schools*, **25** (1), 54–61.

INTERNATIONAL LABOUR OFFICE (1992) *Condition of Work Digest: Preventing Stress at Work (vol. 11, no. 2)*. Geneva: International Labour Office.

INTERNATIONAL LABOUR OFFICE (1993) *World Labour Report 1993*. Geneva: International Labour Office.

IVANCEVICH, J.M. and MATTESON, M.T. (1980) *Stress at Work: A Managerial Perspective.* Glenview, IL: Scott, Foresman and Company.

JACKSON, S.E. (1983) Participation in decision making as a strategy for reducing job-related strain. *Journal of Applied Psychology*, **68**, 3–19.

JACKSON, S.E., SCHWAB, R.L. and SCHULER, R.S. (1986) Toward an understanding of the burnout phenomenon. *Journal of Applied Psychology*, **71**, 630–640.

JACKSON, S.E., TURNER, J.A. and BRIEF, A.P. (1987) Correlates of burnout among public service lawyers. *Journal of Occupational Behaviour*, **8** (4), 339–349.

JAFFE, D.T. and SCOTT, C.D. (1988) *Take this job and love it: How to change your work without changing your job.* New York: Simon & Schuster.

JANSEN, P. (1996) *A Differentiated Practice and Specialisation in Community Nursing* (Doctoral dissertation, Maastricht University). Utrecht: NIVEL.

JASON, L.A., WAGNER, L., TAYLOR, R. *et al.* (1995) Chronic fatigue symdrome: A new challenge for health care professionals. *Journal of Community Psychology*, 23, 143–164.

JENKINS, R. (1991) Introduction. In *Post-viral Fatigue Syndrome* ed. R. JENKINS and J. MOWBRAY, pp. 3–39, Chichester: John Wiley.

JONES, J.W. (1980) *Preliminary Manual: The Staff Burnout Scale for Health Professionals.* Park Ridge, IL: London House.

JUDGE, C.M. (1994) An organisational stress audit: BNR Europe. In *Creating Healthy Work Organisations* ed. C.L. COOPER and S. WILLIAMS, pp. 115–132, Chichester: John Wiley.

KAHILL, S. (1988) Symptoms of professional burnout: A review of the empirical evidence. *Canadian Psychology*, 29, 284–297.

KAHN, R.L. and BYOSIERE, P. (1992) Stress in organizations. In *Handbook of Industrial and Organizational Psychology, Vol. 4* ed. M.D. DUNETTE, J.M.R. HOUGH and H.C. TRIANDIS, pp. 571–650. Palo Alto, CA: Consulting Psychologists Press.

KARASEK, R. (1985) *Job content questionnaire.* Los Angeles, CA: Department of Industrial and Systems Engineering, University of Southern California.

KARGER, H.J. (1981) Burnout as alienation. *Social Service Review*, 55, 270–283.

KASL, S.V. and SEXNER, S. (1992) Health promotion at the worksite. In *International Review of Health Psychology* ed. S. MAES, H. LEVENTHAL and M. JOHNSTON, pp. 111–142, Chichester: John Wiley.

KASLOW, F.W. (1982) Therapy with distressed psychotherapists: Special problems and challenges. In *Professionals in Distress: Issues, Syndromes and Solutions in Psychology* ed. R.R. KILBURG, P.E. NATHAN and R.W. THORESON, pp. 187–210, Washington, DC: American Psychological Association.

KEIJSERS, G.J., SCHAUFELI, W.B., LE BLANC, P.M., ZWERTS, C. and REIS-MIRANDA, D. (1995) Performance and burnout in intensive care units. *Work & Stress*, 9, 513–527.

KELLOWAY, E.K. and BARLING, J. (1991) Job characteristics, role stress and mental health. *Journal of Occupational Psychology*, 64, 291–304.

KERS, W.C. and VAN DER ZOUWE, N. (1994) Psychische ziekten: Volledig en blijvend arbeidsongeschikt? [Mental disorders: Entirely and permanently work incapacitated?] *Tijdschrift voor Sociale Geneeskunde*, 72, 293–300.

KILPATRICK, A.O. (1989) Burnout correlates and validity of research designs in a large panel of studies. *Journal of Health and Human Resources Administration*, 12, 25–45.

KIMMEL, J.M. (1993) Occupational self-evaluations by rehabilitation employees and job burnout. *Dissertation Abstracts International*, 54 (4), 1240A.

KIRK, S.A. and KOESKE, G.F. (1995) The fate of optimism: A longitudinal study of case managers' hopefulness and subsequent morale. *Research on Social Work Practice*, 5, 47–61.

KLEIBER, D. and ENZMANN, D. (1990) *Burnout. Eine internationale Bibliographie – An International Bibliography.* Göttingen: Verlag für Psychologie Dr C.J. Hogrefe.

KLEIBER, D., ENZMANN, D. and GUSY, B. (1995) Stress and burnout in AIDS health care: Are there special characteristics? In *Health Workers and AIDS. Research, Intervention and Current Issues in Burnout and Response* ed. L. BENNETT, D. MILLER and M.W. ROSS, pp. 115–129, Chur (Switzerland): Harwood.

KLEIBER, D., ENZMANN, D. and GUSY, B. (1998) *Skalenhandbuch zur Streß- und Burnoutforschung im medizinisch-psychosozialen Bereich* [Handbook of scales for research on stress and burnout in medical and psychosocial fields]. Göttingen: Verlag für Psychologie Dr C.J. Hogrefe.

KOESKE, G.F. and KOESKE, R.D. (1989) Construct validity of the Maslach Burnout Inventory: A critical review and reconceptualization. *Journal of Applied Behavioral Science*, **25**, 131–144.

KOP, W.J. (1994) *The predictive value of vital exhaustion in the clinical course after coronary angioplasty* (Dissertation). Maastricht: Universitaire Pers Maastricht.

KRAMER, M. (1974) *Reality Shock*. St Louis, MO: Mosby.

KÜNZEL, R. and SCHULTE, D. (1986) 'Burn-out' und Praxisschock Klinischer Psychologen ['Burn-out' and reality shock among clinical psychologists]. *Zeitschrift für Klinische Psychologie, Forschung und Praxis*, **15**, 303–320.

LANDSBERGIS, P.A. (1988) Occupational stress among health care workers: A test of the job demands-control model. *Journal of Organizational Behavior*, **9**, 217–239.

LARSON, D.G. (1986) Developing effective hospice staff support groups: Pilot test of an innovative training program. *Hospice Journal*, **2** (2), 41–55.

LASCH, C. (1979) *The Culture of Narcissism: American Life in an Age of Diminishing Returns*. New York: Norton.

LAWSON, D.A. and O'BRIEN, R.M. (1994) Behavioral and self-report measures of staff burnout in developmental disabilities. *Journal of Organizational Behavior Management*, **14** (2), 37–54.

LAZARO, C., SHINN, M. and ROBINSON, P.E. (1985) Burnout, job performance, and withdrawal behaviors. *Journal of Health and Human Resources Administration*, **7**, 213–234.

LAZARUS, R.S. and FOLKMAN, S. (1984) *Stress, Appraisal, and Coping*. New York: Springer.

LAZARUS, R.S. and LAUNIER, R. (1978) Stress-related transactions between person and environment. In *Perspectives in Interactional Psychology* ed. A. PERVIS and M. LEWIS, pp. 287–323, New York: Plenum Press.

LEAKEY, P., LITTLEWOOD, M., REYNOLDS, S. and BUNCE, D. (1994) Caring for the carers: North Derbyshire Health Authority. In *Creating Healthy Work Organisations* ed. C.L. COOPER and S. WILLIAMS, pp. 167–197, Chichester: John Wiley.

LECROY, C.W. and RANK, M.R. (1986) Factors associated with burnout in the social services: An exploratory study. Special Issue: Burnout among social workers. *Journal of Social Service Research*, **10**, 23–39.

LEE, C. and GRAY, J.A. (1994) The role of employee assistance programmes. In *Creating Healthy Work Organisations* ed. C.L. COOPER and S. WILLIAMS, pp. 215–242, Chichester: John Wiley.

LEE, R.T. and ASHFORTH, B.E. (1990) On the meaning of Maslach's three dimensions of burnout. *Journal of Applied Psychology*, **75** (6), 743–747.

LEE, R.T. and ASHFORTH, B.E. (1993) A longitudinal study of burnout among supervisors and managers: Comparison between the Leiter and Maslach (1988) and Golembiewski *et al.* (1986) models. *Organizational Behavior and Human Decision Processes*, **54**, 369–398.

LEE, R.T. and ASHFORTH, B.E. (1996) A meta-analytic examination of the correlates of the three dimensions of job burnout. *Journal of Applied Psychology*, **81**, 123–133.

LEITER, M.P. (1988) Burnout as a function of communication patterns: A study of a multidisciplinary mental health team. *Group and Organization Studies*, **13**, 111–128.

LEITER, M.P. (1990) The impact of family resources, control coping, and skill utilization on the development of burnout: A longitudinal study. *Human Relations*, **43**, 1067–1083.

LEITER, M.P. (1991) Coping patterns as predictors of burnout: The function of control and escapist coping patterns. *Journal of Organizational Behavior*, **12**, 123–144.

LEITER, M.P. (1992a) Burnout as a crisis in self-efficacy: Conceptual and practical implications. *Work & Stress*, **6**, 107–115.

LEITER, M.P. (1992b) Burnout as a crisis in professional role structures: Measurement and conceptual issues. *Anxiety, Stress, and Coping,* **5** (1), 79–93.

LEITER, M.P. (1993) Burnout as a developmental process: Consideration of models. In *Professional Burnout: Recent Developments in Theory and Research* ed. W.B. SCHAUFELI, C. MASLACH and T. MAREK, pp. 237–250, Washington, DC: Taylor & Francis.

LEITER, M.P. and DURUP, M.J. (1994) The discriminant validity of burnout and depression: A confirmatory factor analytic study. *Anxiety, Stress, and Coping,* **7,** 357–373.

LEITER, M.P. and DURUP, M.J. (1996) Work, home, and in-between: A longitudinal study of spillover. *Journal of Applied Behavioral Science,* **32,** 29–47.

LEITER, M.P. and SCHAUFELI, W.B. (1996) Consistency of the burnout construct across occupations. *Anxiety, Stress, and Coping,* **9,** 229–243.

LEWIN, K. (1945) The Research Center for Group Dynamics at Massachusetts Institute of Technology. *Sociometry,* **2,** 126–136.

LOWMAN, R.L. (1993) *Counseling and Psychotherapy of Work Dysfunctions.* Washington, DC: American Psychological Association.

LUBIN, B., ROBINSON, A.J. and SAILORS, J.R. (1992) Burnout in organizations: A bibliography of the literature, 1980 through 1991. *Organization Development Journal,* **10,** 66–90.

LUNDIN, W. and LUNDING, K. (1994) *The Healing Manager.* New York: Berreta-Kohler.

MAES, S., KITTEL, F., SCHOLTEN, H. and VERHOEVEN, C. (1992) 'Healthier Work at Brabantia', a comprehensive approach to wellness at the worksite. *Safety Science,* **15,** 351–366.

MAHER, E.L. (1983) Burnout and commitment: A theoretical alternative. *Personnel and Guidance Journal,* **61,** 390–393.

MALKINSON, R., KUSHNIR, T. and WEISBERG, E. (1997) Stress management and burnout prevention in female blue-collar workers: theoretical and practical implications. *International Journal of Stress Management,* **4,** 183–197.

MALLETT, K.L., PRICE, J.H., JURS, S.G. and SLENKER, S. (1991) Relationships among burnout, death anxiety, and social support in hospice and critical care nurses. *Psychological Reports,* **68** (Pt 2), 1347–1359.

MASLACH, C. (1982a) Burnout: A social psychological analysis. In *The Burnout Syndrome* ed. J.W. Jones, pp. 30–53, Park Ridge, IL: London House.

MASLACH, C. (1982b) *Burnout. The Cost of Caring.* Englewood Cliffs, NJ: Prentice-Hall.

MASLACH, C. (1984) Personal strategies for managing burnout. In *Employee Assistance Programs in Higher Education* ed. R.W. THORESON and E.P. HOSOKAWA, pp. 159–167, Springfield, IL: Charles C. Thomas.

MASLACH, C. (1993) Burnout: A multidimensional perspective. In *Professional Burnout: Recent Developments in Theory and Research* ed. W.B. SCHAUFELI, C. MASLACH and T. MAREK, pp. 19–32, Washington, DC: Taylor & Francis.

MASLACH, C. and JACKSON, S.E. (1981a) *Maslach Burnout Inventory. Research Edition.* Palo Alto, CA: Consulting Psychologists Press.

MASLACH, C. and JACKSON, S.E. (1981b) The measurement of experienced burnout. *Journal of Occupational Behaviour,* **2,** 99–113.

MASLACH, C. and JACKSON, S.E. (1984a) Burnout in organizational settings. In *Applied Social Psychology Annual* (Vol. 5) ed. S. OSKAMP, pp. 133–153, Beverly Hills, CA: Sage.

MASLACH, C. and JACKSON, S.E. (1984b) Patterns of burnout among a national sample of public contact workers. *Journal of Health and Human Resources Administration,* **7,** 189–212.

MASLACH, C. and JACKSON, S.E. (1985) The role of sex and family variables in burnout. *Sex Roles,* **12,** 837–851.

MASLACH, C. and JACKSON, S.E. (1986) *Maslach Burnout Inventory. Manual* (2nd ed.). Palo Alto, CA: Consulting Psychologists Press.

MASLACH, C., JACKSON, S.E. and LEITER, M. (1996) *Maslach Burnout Inventory. Manual* (3rd ed.). Palo Alto, CA: Consulting Psychologists Press.

MASLACH, C. and LEITER, M.P. (1997) *The Truth about Burnout: How Organizations cause Personal Stress and What to do about it.* San Francisco, CA: Jossey-Bass.

MASLACH, C. and SCHAUFELI, W.B. (1993) Historical and conceptual development of burnout. In *Professional Burnout: Recent Developments in Theory and Research* ed. W.B. SCHAUFELI, C. MASLACH and T. MAREK, pp. 1–16, Washington, DC: Taylor & Francis.

McCRAE, R.R. and JOHN, O.P. (1992) An introduction to the Five Factor model and its applications. *Journal of Personality*, **60**, 175–215.

McDERMOTT, D. (1984) Professional burnout and its relation to job characteristics, satisfaction, and control. *Journal of Human Stress*, **10**, 79–85.

McDONALD, D.G. and HODGDON, J.A. (1991) *Psychological Effect of Aerobic Fitness Training: Research and Theory.* New York: Springer.

McINTOSH, N.J. (1991) Identification and investigation of properties of social support. *Journal of Organizational Behavior*, **12**, 201–217.

McKNIGHT, J.D. (1993) Perceived job control, burnout and depression in hospital nurses: Longitudinal and cross-sectional studies. *Dissertation Abstracts International*, **54** (12), 6499B.

McKNIGHT, J.D. and GLASS, D.C. (1995) Perceptions of control, burnout, and depressive symptomatology: A replication and extension. *Journal of Consulting and Clinical Psychology*, **63**, 490–494.

MEICHENBAUM, D.H. (1985) *Stress inoculation training.* New York: Pergamon Press.

MEIER, S.T. (1983) Toward a theory of burnout. *Human Relations*, **36**, 899–910.

MEIER, S.T. (1984) The construct validity of burnout. *Journal of Occupational Psychology*, **57**, 211–219.

MELAMED, S., KUSHNIR, T. and SHIROM, A. (1992) Burnout and risk factors for cardiovascular diseases. *Behavioral Medicine*, **18**, 53–60.

MELCHIOR, M.E.W., PHILIPSEN, H., ABU-SAAD, H.H. *et al.* (1996) The effectiveness of primary nursing on burnout among psychiatric nurses in long-stay settings. *Journal of Advanced Nursing*, **24**, 694–702.

MELCHIOR, M.E.W., VAN DEN BERG, A.A., HALFENS, R. *et al.* (1997) Burnout and the work-environment of nurses in psychiatric long-stay care settings. *Social Psychiatry and Psychiatric Epidemiology*, **32**, 158–164.

MEYERSON, D.E. (1994) Interpretations of stress in institutions: The cultural production of ambiguity and burnout. *Administrative Science Quarterly*, **39**, 628–653.

MICKLER, S.E. and ROSEN, S. (1994) Burnout in spurned medical caregivers and the impact of job expectancy training. *Journal of Applied Social Psychology*, **24**, 2110–2131.

MILLER, K.I., BIRKHOLT, M., SCOTT, C. and STAGE, C. (1995) Empathy and burnout in human service work: An extension of a communication model. *Communication Research*, **22**, 123–147.

MILLER, K. and KOBELSKI, P. (1982) *Burnout: A Multidisciplinary Bibliography* (Public Administration Series: Bibliography, P-1051). Monticello, IL: Vance Bibliographies.

MILLER, L.S. (1991) The relationship between social support and burnout: Clarification and simplification. *Social Work Research and Abstracts*, **27**, 34–37.

MILLER, T.Q., SMITH, T.W., TURNER, C.W., GUIJARRO, M.L. and HALLET, A.J. (1996) A meta-analytic review of research on hostility and physical health. *Psychological Bulletin*, **119**, 322–348.

MOORE, W.J. (1984) The relationship between unrealistic self-expectations and burnout among pastors. *Dissertation Abstracts International*, **45** (6), 1680A.

MOR, V. and LALIBERTE, L. (1984) Burnout among hospice staff. *Health and Social Work*, **9**, 274–283.

MORRIS, J.A. and FELDMAN, D.C. (1996) The dimensions, antecedents, and consequences of emotional labor. *Academy of Management Review*, **21**, 986–1010.

MOWDAY, R.T., STEERS, R.M. and PORTER, L.W. (1979) The measurement of organizational commitment. *Journal of Vocational Behavior*, **14**, 224–247.

MURPHY, L.R. (1996) Stress management techniques: Secondary prevention of stress. In *Handbook of Work and Health Psychology* ed. M.J. SCHABRACQ, J.A.M. WINNUBST and C.L. COOPER, pp. 427–443, Chichester: John Wiley.

NATIONAL LIBRARY OF AUSTRALIA (ed.) (1981) *Occasional Bibliography No. 3. Occupational Stress and Burnout. A Select Reading List.* Canberra: National Library of Australia.

NORTHWESTERN NATIONAL LIFE INSURANCE COMPANY (1991) *Northwestern National Life survey of working Americans on workplace stress.* Minneapolis, MN: Northwestern National Life Insurance Company.

NOWACK, K.M. (1986) Type A, hardiness, and psychological distress. *Journal of Behavioral Medicine*, **9** (6), 537–548.

NOWACK, K.M. and HANSON, A.L. (1983) The relationship between stress, job performance, and burnout in college student resident assistants. *Journal of College Student Personnel*, **24**, 545–550.

NOWACK, K.M. and PENTKOWSKI, A.M. (1994) Lifestyle habits, substance use and predictors of job burnout in professional working women. *Work & Stress*, **8**, 19–35.

O'LEARY, L. (1993) Mental health at work. *Occupational Health Review*, **45**, 23–26.

OGUS, E.D., GREENGLASS, E.R. and BURKE, R.J. (1990) Gender-role differences, work stress and depersonalization. *Journal of Social Behavior and Personality*, **5**, 387–398.

OSWIN, M. (1978) *Children living in long-stay hospitals.* London: Heinemann Medical.

PAINE, W.S. (1982) The burnout syndrome in context. In *The Burnout Syndrome* ed. J.W. JONES, pp. 1–29, Park Ridge, IL: London House.

PAOLI, P. (1997) *Second European survey on the work environment 1995.* Dublin: European Foundation for the Improvement of Living and Working Conditions, Loughlinstown House.

PARK, R.E. (1934) Industrial fatigue and group morale. *American Journal of Sociology*, **40**, 349–356.

PARKER, P.A. and KULIK, J.A. (1995) Burnout, self- and supervisor-rated job performance, and absenteeism among nurses. *Journal of Behavioral Medicine*, **18**, 581–599.

PARTRIDGE, E. (1961) *A Dictionary of Slang and Unconventional English* (Vol. 1). London: Routledge and Kegan Paul.

PEETERS, M.C.W., BUUNK, B.P. and SCHAUFELI, W.B. (1995) Social interactions and feelings of inferiority among correctional officers: A daily-event recording approach. *Journal of Applied Social Psychology*, **25**, 1073–1089.

PFENNIG, B. and HÜSCH, M. (1994) *Determinanten und Korrelate des Burnout-Syndroms: Eine meta-analytische Betrachtung* [Determinants and correlates of the burnout syndrome: A meta-analytic approach] (Master's Thesis). Berlin: Freie Universität Berlin, Psychologisches Institut.

PICK, D. and LEITER, M.P. (1991) Nurses' perceptions of the nature and causes of burnout: A comparison of self reports and standardized measures. *Canadian Journal of Nursing Research*, **23**, 33–48.

PIEDMONT, R.L. (1993) A longitudinal analysis of burnout in the health care setting: The role of personal dispositions. *Journal of Personality Assessment*, **61**, 457–473.

PIERCE, C.M. and MOLLOY, G.N. (1990) Psychological and biographical differences between secondary school teachers experiencing high and low levels of burnout. *British Journal of Educational Psychology*, **60**, 37–51.

PINES, A.M. (1993) Burnout: An existential perspective. In *Professional Burnout: Recent Developments in Theory and Research* ed. W.B. SCHAUFELI, C. MASLACH and T. MAREK, pp. 33–51, Washington, DC: Taylor & Francis.

PINES, A.M. (1994) The Palestinian 'intifada' and Israelis' burnout. *Journal of Cross Cultural Psychology*, **25**, 438–451.

PINES, A.M. (1996) *Couple burnout: Causes and Cures.* New York: Routledge.

PINES, A.M. and ARONSON, E. (1983) Combatting burnout. *Children and Youth Services Review*, **5**, 263–275.

PINES, A.M. and ARONSON, E. (1988) *Career Burnout: Causes and Cures.* New York: Free Press.

PINES, A.M., ARONSON, E. and KAFRY, D. (1981) *Burnout: From Tedium to Personal Growth.* New York: Free Press.

PINES, A.M. and MASLACH, C. (1978) Characteristics of staff burnout in mental health settings. *Hospital and Community Psychiatry*, **29**, 233–237.

PINES, A.M. and MASLACH, C. (1980) Combating staff burn-out in a day care center: A case study. *Child Care Quarterly*, **9** (1), 5–16.

PLETCHER, P.T. (1987) *An Annotated Bibliography of Current Literature Dealing With Teacher Stress and Burnout: Causes, Effects, and Management* (RIE/Resources in Education, ED289843).

POPE, C.S., TABACHNICK, B.G. and KEITH-SPIEGEL, P. (1995) Ethics in practice: The beliefs and behaviours of psychologists as therapists. In *Ethical Conflicts in Psychology* ed. D.N. BERSOFF, pp. 223–238, Washington DC: American Psychological Association.

POULIN, J.E. and WALTER, C.A. (1993a) Burnout in gerontological social work. *Social Work*, **38**, 305–310.

POULIN, J.E. and WALTER, C.A. (1993b) Social worker burnout: A longitudinal study. *Social Work Research & Abstracts*, **29**, 5–11.

PRICE, L. and SPENCE, S.H. (1994) Burnout symptoms amongst drug and alcohol service employees: Gender differences in the interaction between work and home stressors. *Anxiety, Stress, and Coping*, **7**, 67–84.

PRINS, R. (1990) *Sickness absence in Belgium, Germany (FR) and The Netherlands: A comparative study.* Amsterdam: NIA.

PRÖLL, U. and STREICH, W. (1984) *Arbeitszeit und Arbeitsbedingungen im Krankenhaus* [Work schedule and working conditions in hospitals] (Forschungsbericht Nr. 386). Dortmund: Bundesanstalt für Arbeitsschutz.

QUICK, J.C., PAULUS, P.B., WHITTENGTON, J.L., LARY, T.S. and NELSON, D.L. (1996) Management development, well-being and health. In *Handbook of Work and Health Psychology* ed. M.J. SCHABRACQ, J.A.M. WINNUBST and C.L. COOPER, pp. 369–388, Chichester: John Wiley.

RABINOWITZ, S., KUSNIR, T. and RIBAK, J. (1996) Preventing burnout: Increasing professional self-efficacy in primary care nurses in a Balint group. *American Association of Occupational Health Nurses*, **44**, 28–32.

RAFFERTY, J.P., LEMKAU, J.P., PURDY, R.R. and RUDISILL, J.R. (1986) Validity of the Maslach Burnout Inventory for family practice physicians. *Journal of Clinical Psychology*, **42**, 488–492.

RANDALL, M. and SCOTT, W.A. (1988) Burnout, job satisfaction, and job performance. *Australian Psychologist*, **23**, 335–347.

RANDOLPH, S. (1981) Peer support groups: The Ohio model for alleviating staff burnout. In *The Burnout Syndrome: Current Research, Theory, Interventions* ed. J.W. JONES, pp. 165–171, Park Ridge, IL: London House.

RAQUEPAW, J.M. and MILLER, R.S. (1989) Psychotherapist burnout: A componential analysis. *Professional Psychology: Research and Practice*, **20**, 32–36.

READ, K.E. (1987) *An Annotated Bibliography Concerning Teacher Stress and Burnout: Causes and Management Techniques* (RIE/Resources in Education, ED289844).

REILLY, N.P. (1994) Exploring a paradox: Commitment as a moderator of the stressor-burnout relationship. *Journal of Applied Social Psychology*, **24**, 397–414.

RIFKIN, J. (1995) *The End of Work*. New York: Tracher/Putnam.

RIGGAR, T.F. (1985) *Stress Burnout. An Annotated Bibliography*. Carbondale and Edwardsville, IL: Southern Illinois University Press.

RIGGAR, T.F. and BEARDSLEY, M. (1983) *Stress-Burnout: A Bibliography. 1000+ References* (RIE/Resources in Education, ED240411).

ROBINSON, S.L. and ROUSSEAU, D.M. (1994) Violating the psychological contract: Not the expectation but the norm. *Journal of Organizational Behavior*, **15**, 245–259.

ROSS, R.R. and ALTMAIER, E.M. (1994) *Interventions in Occupational Stress*. London: Sage.

ROSSE, J.G., BOSS, R.W., JOHNSON, A.E. and CROWN, D.F. (1991) Conceptualizing the role of self-esteem in the burnout process. *Group and Organization Studies*, **16**, 428–451.

ROUNTREE, B.H. (1984) Psychological burnout in tasks groups: Examining the proposition that some task groups of workers have an affinity for burnout, while others do not. *Journal of Health and Human Resources Administration*, **7**, 235–248.

ROUSSEAU, D.M. (1989) Psychological and implied contracts in organizations. *Employee Responsibilities and Rights Journal*, **2**, 212–139.

ROUSSEAU, D.M. and PARKES, J.M. (1993) The contracts of individuals and organizations. In *Research in Organizational Behavior* (Vol. 15) ed. L.L. CUMMINGS and B.M. STAW, pp. 1–43, Greenwich, CT: JAI Press.

ROYAL COLLEGES OF PHYSICIANS, PSYCHIATRISTS AND GENERAL PRACTITIONERS (1996) *Chronic fatigue syndrome* (Report of a joint working group of the Royal Colleges of Physicians, Psychiatrists and General Practitioners).

SALERNO, A.J. (1991) Work expectations and burnout in mental health professionals. *Dissertation Abstracts International*, **52** (6), 3331B.

SARROS, J.C. and FRIESEN, D. (1987) The etiology of administrator burnout. *Alberta Journal of Educational Research*, **33**, 163–179.

SATYAMURTI, C. (1981) *Occupational Survival*. Oxford: Blackwell.

SCHAAP, C., SCHAUFELI, W.B. and HOOGDUIN, C. (1995) Diagnostiek en behandeling van chronische werkstress en burnout [Diagnosis and treatment of chronic job stress and burnout]. *Directieve Therapie*, **15**, 215–232.

SCHAUFELI, W.B. (1995) The evaluation of a burnout workshop for community nurses. *Journal of Health and Human Services Administration*, **18**, 11–31.

SCHAUFELI, W.B., DAAMEN, J. and VAN MIERLO, H. (1994a) Burnout among Dutch teachers: An MBI-validity study. *Psychological and Educational Measurement*, **54**, 803–812.

SCHAUFELI, W.B., ENZMANN, D. and GIRAULT, N. (1993a) Measurement of burnout: A review. In *Professional Burnout: Recent Developments in Theory and Research* ed. W.B. SCHAUFELI, C. MASLACH and T. MAREK, pp. 199–215, Washington, DC: Taylor & Francis.

SCHAUFELI, W.B., HOOGDUIN, C.A.L., SCHAAP, C. and KLADLER, A. (in press). *Burnout in an outpatient sample: The clinical validity of the Maslach Burnout Inventory and the Burnout Measure* (Unpublished manuscript. Psychological Department, Utrecht University).

SCHAUFELI, W.B. and JANCZUR, B. (1994) Burnout among nurses. A Polish-Dutch comparison. *Journal of Cross-Cultural Psychology*, **25**, 95–113.

SCHAUFELI, W.B., KEIJSERS, G.J. and REIS-MIRANDA, D. (1995) Technology use, burnout, and performance in Intensive Care Units. In *Organizational Risk Factors for Job Stress* ed. S.L. SAUTER and L.R. MURPHY, pp. 259–271, Washington, DC: American Psychological Association.

SCHAUFELI, W.B. and LE BLANC, P. (1997) Personnel. In *Organisation and Management of Intensive Care: A Prospective Study in 12 European Countries* ed. D. REIS MIRANDA, D.W. RYAN, W.B. SCHAUFELI and V. FIDLER, pp. 169–208, Berlin: Springer.

SCHAUFELI, W.B., LEITER, M.P., MASLACH, C. and JACKSON, S.E. (1996a) The MBI-General Survey. In *Maslach Burnout Inventory* (3rd edn.) ed. C. MASLACH, S.E. JACKSON and M. LEITER, pp. 19–26, Palo Alto, CA: Consulting Psychologists Press.

SCHAUFELI, W.B., MASLACH, C. and MAREK, T. (eds) (1993b) *Professional Burnout: Recent Developments in Theory and Research*. Washington, DC: Taylor & Francis.

SCHAUFELI, W.B., VAN DEN EYNDEN, R.J.J.M. and BROUWERS, H.M.G. (1994b) Stress en burnout bij penitentiare inrichtingswerkers: De rol van sociaal-cognitieve factoren [Stress and burnout among correctional officers: The role of social-cognitive factors]. *Gedrag & Organisatie*, **7**, 216–224.

SCHAUFELI, W.B. and VAN DIERENDONCK, D. (1993) The construct validity of two burnout measures. *Journal of Organizational Behavior*, **14**, 631–647.

SCHAUFELI, W.B. and VAN DIERENDONCK, D. (1994) Burnout, een begrip gemeten. De Nederlandse versie van de Maslach Burnout Inventory (MBI-NL) [Burnout; the measurement of a concept: The Dutch version of the MBI]. *Gedrag & Gezondheid*, **22**, 153–172.

SCHAUFELI, W.B. and VAN DIERENDONCK, D. (1995) A cautionary note about the cross-national and clinical validity of cut-off points for the Maslach Burnout Inventory. *Psychological Reports*, **76**, 1083–1090.

SCHAUFELI, W.B., VAN DIERENDONCK, D. and VAN GORP, K. (1996b) Burnout and reciprocity: Towards a dual-level social exchange model. *Work & Stress*, **3**, 225–237.

SCHREURS, P.J.G., WINNUBST, J.A.M. and COOPER, C.L. (1996) Workplace health programmes. In *Handbook of Work and Health Psychology* ed. M.J. SCHABRACQ, J.A.M. WINNUBST and C. L. COOPER, pp. 463–481, Chichester: John Wiley.

SCHROËR, K. (1993) *Verzuim wegens overspanning* [Absenteeism because of burnout]. Maastricht: Universitaire Pers Maastricht.

SCHWAB, R.L., JACKSON, S.E. and SCHULER, R.S. (1986) Educator burnout: Sources and consequences. *Educational Research Quarterly*, **10** (3), 14–30.

SCHWARTZ, M.S. and WILL, G.T. (1953) Low morale and mutual withdrawal on a mental hospital ward. *Psychiatry*, **16**, 337–353.

SEABOLD, D.R. (1983) The development of a predictive model and measure of occupational burnout in the nursing profession. *Dissertation Abstracts International*, **45**, 1595B.

SEMMER, N. (1996) Individual differences, work stress and health. In *Handbook of Work and Health Psychology* ed. M.J. SCHABRACQ, J.A.M. WINNUBST and C.L. COOPER, pp. 51–86, Chichester: John Wiley.

SHAPIRO, D.A., BARKHAM, M., HARDY, G.E. and MORRISON, L.A. (1990) The second Sheffield Psychotherapy Project: Rationale, design and preliminary outcome data. *British Journal of Medical Psychology*, **63**, 97–108.

SHAPIRO, D.A. and FIRTH, J.A. (1987) Prescriptive vs. exploratory psychotherapy: Outcomes of the Sheffield Psychotherapy Project addendum. *British Journal of Psychiatry*, **151**, 790–799.

SHEPHARD, R.J. (1996) Financial aspects of employee fitness programmes. In *Workplace Health: Employee Fitness and Exercise* ed. J. KERR, A. GRIFFITHS and T. COX, pp. 29–54, London: Taylor & Francis.

SHIROM, A. (1989) Burnout in work organizations. In *International Review of Industrial and Organizational Psychology 1989* ed. C.L. COOPER and I.T. ROBERTSON, pp. 25–48, Chichester: John Wiley.

SHIROM, A. and OLIVER, A. (1986) *Does stress lead to affective strain, or vice versa? A cross-lagged regression test* (Paper presented at the 21st Congress of the International Association of Applied Psychology, July 1986, Jerusalem, Israel).

SHIROM, A., WESTMAN, M., SHAMAI, O. and CAREL, R.S. (1997) Effects of work overload and burnout on cholesterol and triglycerides levels: The moderating effects of employee reactivity among male and female employees. *Journal of Occupational Health Psychology*, **2**, 275–288.

SIEFERT, K., JAYARATNE, S.D. and CHESS, W.A. (1991) Job satisfaction, burnout, and turnover in health care social workers. *Health and Social Work*, **16**, 193–202.

SIEGRIST, J. (1996) Adverse health effects of high-effort/low-reward conditions. *Journal of Occupational Health Psychology*, **1**, 27–41.

SIXMA, H.J., BAKKER, A.B., BOSVELD, W. *et al.* (1998) *Dropping out as a consequence of burnout: A longitudinal study among Dutch general practitioners* (Unpublished manuscript. Psychological Department, Utrecht University).

SLATER, P. (1976) *The Pursuit of Loneliness* (Rev. edn.). Boston: Beacon.

SÖDERFELDT, M., SÖDERFELDT, B., WARG, L.E. and OHLSON, C.-G. (1996) The factor structure of the Maslach Burnout Inventory in two Swedish human service organizations. *Scandinavian Journal of Psychology*, **37**, 437–443.

STARNAMAN, S.M. and MILLER, K.I. (1992) A test of a causal model of communication and burnout in the teaching profession. *Communication Education*, **41** (1), 40–53.

STEVENS, G.B. and O'NEILL, P. (1983) Expectation and burnout in the developmental disabilities field. *American Journal of Community Psychology*, **11**, 615–627.

STOUT, J.K. and WILLIAMS, J.M. (1983) Comparison of two measures of burnout. *Psychological Reports*, **53**, 283–289.

STROEBE, W. and STROEBE, M. (1991) Partnerschaft, Familie und Wohlbefinden [Partnership, family, and well-being]. In *Wohlbefinden: Theorie, Empirie, Diagnostik* [Well-being: Theory, research, diagnostics] ed. A. ABELE and P. BECKER, pp. 155–174, Weinheim: Juventa.

SUTHERLAND, V.J. and COOPER, C.L. (1990) *Understanding Stress: A Psychological Perspective for Health Professionals.* London: Chapman & Hall.

TAYLOR, H. and COOPER, C.L. (1989) The stress prone personality: A review of research in the context of occupational stress. *Stress & Medicine*, **5**, 17–27.

THARENOU, P. (1979) Employee self-esteem: A review of the literature. *Journal of Vocational Behavior*, **15**, 316–346.

THE LANCET (1994) Burnished or burnt out: the delights and dangers of working in health (editorial). *The Lancet*, **344** (8937), 1583–1584.

TUKE, D.H. (1882) *A dictionary of psychological medicine.* New York: Arno Press.

URSPRUNG, A.W. (1986) Incidence and correlates of burnout in residential service settings. *Rehabilitation Counseling Bulletin*, **29**, 225–239.

US DEPARTMENT OF LABOR (1990) *What works: Workplaces without Drugs.* Washington, DC: US Department of Labor.

VAN DER KLINK, J.J.L. and TERLUIN, B. (1996) Begeleiding en interventies bij overspanning in de eerste lijn [Burnout counselling and interventions in primary health care].

In *Behandelingsstrategieën bij burnout* ed. C.A.L. HOOGDUIN, C.P.D.R. SCHAAP, A.J. KLADLER and W.A. HOOGDUIN, pp. 21–32, Houten: Bohn Stafleu van Loghum.

VAN DIERENDONCK, D., SCHAUFELI, W.B. and BUUNK, B.P. (1996) Inequality among human service professionals: Measurement and relation to burnout. *Basic and Applied Social Psychology*, **18**, 429–451.

VAN DIERENDONCK, D., SCHAUFELI, W.B. and BUUNK, B.P. (in press). The evaluation of an individual burnout program. The role of inequity and social support. *Journal of Applied Psychology*.

VAN DIERENDONCK, D., SCHAUFELI, W.B. and SIXMA, H.J. (1994) Burnout among general practitioners: A perspective from equity theory. *Journal of Social and Clinical Psychology*, **13**, 86–100.

VAN GORP, K. and SCHAUFELI, W.B. (1996) *Een gezonde geest in een gezond lichaam* [A healthy mind in a healthy organisation]. The Hague: VUGA.

VAN GORP, K., SCHAUFELI, W.B. and HOPSTAKEN, L. (1993) Burnout en cognitief terugtrekgedrag vanuit sociaal uitwisselingsperspectief [Burnout and cognitive withdrawal from a social exchange perspective]. *Gedrag en Gezondheid*, **21**, 274–285.

VAN HORN, J.E., SCHAUFELI, W.B. and ENZMANN, D. (in press) Teacher burnout and lack of reciprocity. *Basic and Applied Social Psychology*.

VAN HORN, J.E., SCHAUFELI, W.B., GREENGLASS, E.R. and BURKE, R.J. (1997) A Canadian-Dutch comparison of teachers' burnout. *Psychological Reports*, **81**, 371–382.

VAN YPEREN, N.W. (1996) Communal orientation and the burnout syndrome among nurses: A replication and extension. *Journal of Applied Social Psychology*, **26**, 338–354.

VAN YPEREN, N.W., BUUNK, B.P. and SCHAUFELI, W.B. (1992) Communal orientation and the burnout syndrome among nurses. *Journal of Applied Social Psychology*, **22**, 173–189.

VANDENBOS, G.R. and DUTHIE, R.F. (1986) Confronting and supporting colleagues in distress. In *Professionals in Distress: Issues, Syndromes and Solutions in Psychology* ed. R.R. KILBURG, P.E. NATHAN and R.W. THORESON, pp. 211–232, Washington, DC: American Psychological Association.

WADE, D.C., COOLEY, E.J. and SAVICKI, V. (1986) A longitudinal study of burnout. *Children and Youth Services Review*, **8**, 161–173.

WALLACE, J.E. and BRINKEROFF, M.B. (1991) The measurement of burnout revisited. *Journal of Social Service Research*, **14**, 85–111.

WARR, P. (1987) *Work, Unemployment and Mental Health*. Oxford: Oxford University Press.

WEINBERG, S., EDWARDS, G.M. and GAROVE, W.E. (1983) Burnout among employees of state residential facilities serving developmentally disabled persons. *Children and Youth Services Review*, **5**, 239–253.

WEINER, M.F., CALDWELL, T. and TYSON, J. (1983) Stress and coping in ICU nursing: Why support groups fail. *General Hospital Psychiatry*, **5**, 179–183.

WEST, D.J., HORAN, J.J. and GAMES, P.A. (1984) Component analysis of occupational stress inoculation applied to registered nurses in an acute care hospital setting. *Journal of Counseling Psychology*, **31**, 208–218.

WESTMAN, M. and EDEN, D. (1997) Effects of a respite from work on burnout: Vacation relief and fade-out. *Journal of Applied Psychology*, **82**, 516–527.

WESTMAN, M. and ETZION, D. (1995) Crossover of stress, strain and resources from one spouse to another. *Journal of Organizational Behavior*, **16**, 169–181.

WHEATON, B. (1996) The domains and boundaries of stress concepts. In *Psychosocial Stress: Perspectives on Structure, Theory, Life-course and Methods* ed. H.B. KAPLAN, pp. 29–70, New York: Academic Press.

WOLPIN, J., BURKE, R.J. and GREENGLASS, E.R. (1991) Is job satisfaction an antecedent or a consequence of psychological burnout? *Human Relations*, **44**, 193–209.

WORLD HEALTH ORGANIZATION (1992) *The ICD-10 Classification of Mental and Behavioural Disorders: Clinical Description and Diagnostic Guidelines.* Geneva: World Health Organization.

ZEDECK, S., MASLACH, C., MOSIER, K. and SKITKA, L. (1988) Affective response to work and quality of family life: Employee and spouse perspectives. *Journal of Social Behavior and Personality*, **3**, 135–157.

ZIMBARDO, P.G. (1970) The human choice: Individuation, reason, and order versus determination, impulse and chaos. In *Nebraska Symposium in Motivation, 1969* ed. W.J. ARNOLD and D. LEVINE, pp. 237–307, Lincoln, NB: University of Nebraska Press.

ZWERTS, C., SCHAUFELI, W.B., KEIJSERS, G.J., LE BLANC, P.M. and REIS MIRANDA, D. (1995) Burnout en prestatie in Intensive Care Units. [Burnout and performance in Intensive Care Units] *Tijdschrift voor Sociale Gezondheidszorg*, **73**, 382–389.

Index